Computational and Mathematical Models of Microstructural Evolution

MATERIALS RESEARCH SOCIETY
SYMPOSIUM PROCEEDINGS VOLUME 529

Computational and Mathematical Models of Microstructural Evolution

Symposium held April 13–17, 1998, San Francisco, California, U.S.A.

EDITORS:

Jeffrey W. Bullard
University of Illinois at Urbana-Champaign
Urbana, Illinois, U.S.A.

Long-Qing Chen
The Pennsylvania State University
University Park, Pennsylvania, U.S.A.

Rajiv K. Kalia
Louisiana State University
Baton Rouge, Louisiana, U.S.A.

A. Marshall Stoneham
University College London
London, United Kingdom

Materials Research Society
Warrendale, Pennsylvania

Single article reprints from this publication are available through
University Microfilms Inc., 300 North Zeeb Road, Ann Arbor, Michigan 48106

CODEN: MRSPDH

Published by:

Materials Research Society
506 Keystone Drive
Warrendale, PA   15086
Telephone (724) 779-3003
Fax (724) 779-8313
Website: http://www.mrs.org/

Library of Congress Cataloging in Publication Data

Computational and mathematical models of microstructural evolution :
    symposium held April 13–17, 1998, San Francisco, California, U.S.A. /
    editors, Jeffrey W. Bullard, Long-Qing Chen, Rajiv K. Kalia, A. Marshall
    Stoneham
        p. cm.—(Materials Research Society symposium proceedings,
        ISSN 0272-9172 ; v. 529)
        Includes bibliographical references and index.
        ISBN 1-55899-435-1
        1. Microstructure—Mathematical models. I. Bullard, Jeffrey W. II. Chen,
    Long-Qing III. Kalia, Rajiv K. IV. Stoneham, A. Marshall V. Series: Materials
    Research Society symposium proceedings ; v. 529
TA407.C64   1998   98-44416
620.1'1299—dc21   CIP

Manufactured in the United States of America

CONTENTS

*Invited Paper

PART III: <u>OTHER APPLICATIONS OF</u>
<u>MATHEMATICAL MODELLING</u>

*Invited Paper

PREFACE

This volume contains papers presented at the symposium on "Computational and Mathematical Models of Microstructural Evolution," held April 13–17 at the 1998 MRS Spring Meeting in San Francisco, California. The three-day symposium was designed to bring together the foremost materials theorists and applied mathematicians from around the world to share and discuss some of the newest and most promising mathematical and computational tools for simulating, understanding, and predicting the various complex processes that occur during the evolution of microstructures. The proceedings contains 26 peer-reviewed papers, drawn from the 43 oral presentations—including 11 invited talks—and 30 posters.

We extend our gratitude to the session chairs for ensuring the smooth progression of the symposium. We also wish to thank all of the individual authors and referees for helping with the preparation of the proceedings. Thanks are also due to the MRS staff for all of their assistance in the organization and administration of the symposium, and to Carey Shuey in the Ceramics Program at Penn State for her help in putting together the proceedings. Finally, we wish to gratefully acknowledge the support of the Center for Theoretical and Computational Materials Science at the National Institute of Standards and Technology, without which it would not have been possible to assemble such an outstanding international group of scientists for this symposium.

Jeffrey W. Bullard
Long-Qing Chen
Rajiv K. Kalia
A. Marshall Stoneham

September 1998

MATERIALS RESEARCH SOCIETY SYMPOSIUM PROCEEDINGS

MATERIALS RESEARCH SOCIETY SYMPOSIUM PROCEEDINGS

Prior Materials Research Society Symposium Proceedings available by contacting Materials Research Society

Part I

Mathematics of Microstructure Stability and Evolution

MICROSTRUCTURAL EVOLUTION AND METASTABILITY IN ACTIVE MATERIALS

DAVID KINDERLEHRER
Department of Mathematical Sciences
Carnegie Mellon University, Pittsburgh, PA 15213-3890

ABSTRACT

Metastable systems pose significant problems in both analysis and simulation. We discuss here the evolution of microstructure in a shape-memory alloy where energetic contributions from disparate scales play determining roles. This is a challenge for modelling since the finest length scales cannot be 'seen' at macroscopic level. We then provide a mechanism for kinetics that gives a different notion metastability.

INTRODUCTION

We address here modelling the evolution of microstructure in a mosaic of twinned and compound twinned lamellar structures in shape memory CuAlNi, Abeyartne, Chu and James [1]. In this system there is an alternation between two martensitic variants under loading in which the material selects a succession of fine phase intermediate configurations. At each value of the load, the system is in equilibrium save for small regions near mosaic interface, and, during the process, one variant grows at the expense of the other. Although the entire system has the appearance of a succession of equilibria, there is hysteresis in the macroscopic volume fractions of the variants. Thus the system traverses a succession of metastable states. The origin of this behavior is the small amount of energy stored in the mosaic interfaces. These are, however, spatially localized. So this system exhibits metastable behavior across disparate length scales and is an excellent candidate for methods of multiscale analysis. This is a report on joint work with Richard Jordan and Felix Otto, [2],[3],[4],[5].

A typical problem of multiscale analysis proceeds by the introduction of a parameter in a continuum or phenomenological description of the system under discussion and subsequent passage to the limit as this parameter tends to zero or to infinity. Or, as a sort of inverse to this approach, the behavior of the system at a small scale may be assumed and a limit process analogous to an infinite parallel circuit attempted to obtain an effective macroscopic description, [6]. These methods are very useful when the physics of the system is in the realm of continuum mechanics at all the scales and, importantly, when a clear idea of the energy of the system is available for its complete operating range. For example the effective stress tensor of a polycrystalline material may be obtained by such considerations, but these methods begin to fail when properties involving the grain boundaries are sought. A chemical system involving huge numbers of particles behaving in some random fashion, according to a Langevin Equation, may be understood at macroscopic level by means of its distribution function, the solution of a Fokker-Planck Equation. In this case, the nature of the physics itself has changed, since concern is now directed to evolution of the distribution function. Even here, however, there is little hint about the appropriate dynamics of the system at the macroscopic scale, since knowledge of the state of the system at each time does not of itself provide a way to identify nearby states.

Here we wish to consider a somewhat different approach, involving several coarse graining mechanisms, and seek to shed some insight on [1]. The principal device we employ is a representation of local spatial averages of a configuration in terms of a distribution, a measure called the Young Measure [7],[8]. Although a configuration of a nonlinear material may appear macroscopically as nearly homogeneous, it may consist on interrogation to have complex microstructure, for example of arrangements of fine phase laminates or defect structures. The Young Measure is especially useful in this situation, cf. [8] and [7],[6] for particular applications. Our interpretation of the kinetics of the system will be in terms of how this measure changes and leads to a new interpretation of its metastability.

1. CONSTRAINED THEORY AND COARSE GRAINING

The macroscopic energy of a configuration of the material subjected to a dead load is given by

$$E = \int_\Omega (W(\nabla y) - S \cdot \nabla y) \, dx ,$$ (1.1)

where $y: \Omega \to \mathbf{R}^n$ ($n = 2$ or 3) is the deformation, W is the Helmholtz free energy, and S is the load. The key feature of this energy W is that the symmetries of the lattice imply it has potential wells: W is minimized on a set of copies of $SO(3)$, e.g., $\Sigma = SO(3)U_1 \cup \dots \cup SO(3)U_N$, where the U_k are explicit matrices that emerge from theory, [7],[9],[10]. Equilibrium configurations are sought as minimizers of (1.1). Neglected at this scale, for example, are surface energies, higher gradients, or energies of internal boundaries. The energy is lumpy and rough owing to the complex crystallography of the material. When equilibria are sought, minimizing sequences tend to populate several wells leading to highly oscillatory sequences that converge only weakly.

We turn our attention to the distribution of this population of values, which gives rise to the Young measure. In this context, our measure is a family $\nu = (\nu_x)_{x \in \Omega}$ of probability measures generated by a sequence of deformations

$$y^k: \Omega \to \mathbf{R}^n, \ \nabla y^k \ \overset{*}{\longrightarrow} \ \nabla y \text{ in } L^\infty, \text{ such that for any continuous } f,$$
$$f(\nabla y^k) \ \overset{*}{\longrightarrow} \ \bar{f} \text{ in } L^\infty, \ \bar{f}(x) = \int_M f(A) \, d\nu_x(A) \text{ in } \Omega \text{ a.e.,}$$ (1.2)

where $\mathbf{M}$ denotes matrices. We may interpret y^k as a snapshot of the deformation at scale proportional to $1/k$. The energy and deformation gradient of a configuration subjected to a constant dead load S now has the expression

$$E = \int_\Omega \int_M (W(A) - S \cdot A) \, d\nu_x(A) \text{ and } \nabla y(x) = \int_M A \, d\nu_x(A).$$ (1.3)

The constrained theory consists in identifying a set of admissible measures for the system, typically a subset of ν with supp $\nu \subset \Sigma$. In CuAlNi there are six potential wells comprising this set, they are described in [1],[7],[9]. In our model problem, a further constraint arises by optimizing E over lamellar structures which populate two wells, say $SO(3)U_1$ and $SO(3)U_2$ subject to a biaxial load S with $Se_3 = 0$. The wells U_1 and U_2 have e_3 as a common axis perpendicular to the plane of the sample, so from this point on, we think of our system as being two dimensional. The result of this process is that we may restrict our attention to a one parameter family and homogeneous deformation gradient

$$\nu^{(\xi)} = (1-\xi)\delta_{M_1(\xi)} + \xi\delta_{M_2(\xi)}, \ M_i(\xi) \in SO(3)U_i, \ 0 \le \xi \le 1,$$ (1.4)
$$F(\xi) = (1-\xi)M_1(\xi) + \xi M_2(\xi), \ M_2(\xi) - M_1(\xi) = a(\xi) \otimes n, \ |n| = 1,$$ (1.5)

4

and subject to the fine phase coherence property (rank-one condition) with respect to the second laminate system where it encounters U_2,

$$F(\xi) = U_2 (1 + b(\xi) \otimes m(\xi)), \ |m(\xi)| = 1. \tag{1.6}$$

The energy per unit area at ξ is given by a function

$$E_{loading} = \psi_{macro}(\xi, S) = -\int_M S \cdot A \ dv_x(A) = -F(\xi) \cdot S, \ 0 \le \xi \le 1, \tag{1.7}$$

which is nearly linear, or even concave, [1]. This is the macroscopic contribution, but note that owing to the optimization process, which involves varying over rotations, it is a function of the Young measure through the volume fraction ξ.

The second and third contributions to the energy cannot be seen by the average bulk deformation $F(\xi)$ because they are regulated by finer scales. However, we can express them in terms of the Young Measure. Here we summarize our results. In the mesoscale regime, there is some stress near the interface between the fine phase laminate given by $F(\xi)$ and U_2, say $x \cdot m(\xi) = 0$, because it is fine phase coherent in the sense that (1.6) holds, but it is not exactly coherent in the sense that $M_1(\xi) - U_2 \ne$ rank 1 and $M_2(\xi) - U_2 \ne$ rank 1. This term may be written, after some calculation for an appropriate f in (1.2),

$$E_{transition \ layer} = \psi_{meso}(\xi) = c(\xi - \frac{1}{2})^2 + c_0, \tag{1.8}$$

The contributions (1.7) and (1.8) are derived in [1].

Fine scale oscillations are implicated in the dynamical behavior of the sample and this will account for the "wiggles" in the analysis of [1]. We suggest that their origin arises from small distortions of the lattice near the interface $x \cdot m(\xi) = 0$. By idealizing this situation as a regular lattice to one side of a line, cut at a prescribed slope α, we may obtain a crude notion of the nature of this contribution. (1.6) implies that α is proportional to ξ. In general, the format of embedded atom potentials may be applied, [11],[12]. We obtain here, where δ and h are lattice parameters, $1/K$ is a scale factor, and $\langle \rangle$ denotes a periodic function of period 1,

$$\psi_{micro}(\xi) = \Psi(\frac{1}{K} \sum_1^K f(\delta \langle \frac{\alpha h}{\delta} k \rangle)) \tag{1.9}$$

For a special choice of Ψ and f,

$$\psi_{micro}(\xi) = \varepsilon \ \psi_0(\frac{\xi}{\varepsilon}), \quad \varepsilon \approx 1/K, \tag{1.10}$$

which is the ansatz of [1]. We thus arrive at an idealized effective energy per unit interface length of the form

5

$$\psi_\varepsilon(\xi) \;=\; \psi_{macro}(\xi, S) \;+\; \psi_{meso}(\xi) \;+\; \varepsilon\,\psi_o\!\left(\frac{\xi}{\varepsilon}\right). \qquad (1.11)$$

2. EVOLUTION OF THE MICROSTRUCTURE AND FOKKER-PLANCK DYNAMICS

The sample is subjected to a loading program $S = S(t)$. The most straight forward assumption about its motion is the driving force equation

$$\frac{d\xi}{dt} \;=\; -\mu\,\frac{\partial\psi_\varepsilon}{\partial\xi}, \quad t > 0. \qquad (2.1)$$

(We take $\mu = 1$ for convenience in the sequel.) To solve this equation iteratively, e.g., by backward-Euler, given $\xi^{(k-1)}$, determine $\xi^{(k)}$ the solution of

$$\frac{1}{\tau}(\xi - \xi^{(k-1)}) \;=\; -\psi_\varepsilon'(\xi),$$

which is the same as asking for the ξ such that

$$\frac{1}{2\tau}(\xi - \xi^{(k-1)})^2 \;+\; \psi_\varepsilon(\xi) \;=\; \min. \qquad (2.2)$$

The above is an expression of competition between the distance of the nearby states $\xi^{(k-1)}$, $\xi^{(k)}$ and the energy $\psi_\varepsilon(\xi^{(k)})$. This suggests a second coarse graining in terms of the distributions of the volume fractions. Given a distribution $\rho^{(k-1)}$ of $\xi^{(k-1)}$ we should seek a distribution $\rho^{(k)}$ of $\xi^{(k)}$ in such a way that

$$\frac{1}{2\tau}\int_R\int_R (\xi - \eta)^2\, dp(\xi,\eta) \;+\; \int_R \psi_\varepsilon\, \rho\, d\xi \;+\; \sigma \int_R \rho\log\rho\, d\xi \;=\; \min \qquad (2.3),$$

where p is a joint distribution of $\rho^{(k-1)}$ and ρ and the last term represents a configurational entropy. The first term above is related to the Wasserstein metric and defines a weak toplogy on the distributions $f(\xi)\, d\xi$. The second and third terms represent the free energy of the system. This determines the kinetics and from it we understand metastability as a competition between distance and energy. We may calculate and Euler Equation for this process and it turns out that as $\tau \to 0$, the sequence of solutions $(\rho^{(k)})$ converges to the solution of the Fokker-Planck Equation

$$\frac{\partial\rho}{\partial t} \;=\; \sigma\,\frac{\partial^2\rho}{\partial\xi^2} \;+\; \frac{\partial}{\partial\xi}(\psi_\varepsilon'\rho), \quad -\infty < \xi < \infty,\, t > 0, \qquad (2.4)$$

We may resolve the path of the motion by computing the solution of (2.4), either analytically or by a numerical technique of our choice.

3. SAMPLE SIMULATION: A CREEP TEST

Consider a creep test. Here the material is held in a position, loaded, and then released. Set

$$\psi(\xi) \;=\; \frac{1}{2}\,(\xi - \frac{1}{2})^2 \quad\text{and}\quad \psi_\varepsilon(\xi) \;=\; \frac{1}{2}\,(\xi - \frac{1}{2})^2 \;+\; \varepsilon\psi_o\!\left(\frac{\xi}{\varepsilon}\right),\ |\,\psi_0'\,| \le a. \qquad (3.1)$$

6

In the creep test, we solve (2.4) for ψ and ψ_ε with ρ_0 concentrated near $\xi = 0$. We used an explicit finite difference scheme to calculate the solution. The averages

$$\langle \xi \rangle_t = \int_R \xi \rho \, d\xi \quad \text{and} \quad \langle \xi_\varepsilon \rangle_t = \int_R \xi \rho_\varepsilon \, d\xi \tag{3.2}$$

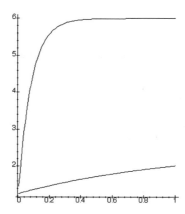

Fig. 1. The results of a creep test, showing the mean values $\langle \xi \rangle_t$, upper curve, and $\langle \xi_\varepsilon \rangle_t$, lower curve. $\langle \xi \rangle_t$ has reached its stationary state by $t = 0.45$.

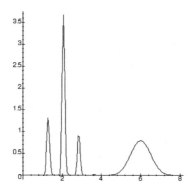

Fig. 2. Final distributions ρ_ε, on the left, and ρ, on the right at $t = 1$ of (2.4).

7

should display the quite different behavior: $\langle \xi \rangle_t$ should saturate near its stationary value $\frac{1}{2}$ quickly while $\langle \xi_\varepsilon \rangle_t$ should advance slowly. This is indeed the case in the example shown in Figure 1, (which was computed, however, with a scale for ξ whose saturation value is 6). The densities ρ and ρ_ε at the end of the simulation are shown in Figure 2. ρ in fact is essentially the Gibbs distribution. Hence, we believe that Fokker-Planck dynamics can describe metastable systems governed by competition in weak topologies. There are, of course, other dynamical mechanisms that describe other situations.

CONCLUSIONS

We have reviewed a metastable system characterized by a reversible transforming microstructure with

(a)   a heirarchy of scales analysis of energetics based on a coarse graining that employs statistics of the deformation (the Young Measure or other device) and

(b)   the interpretation of metastable evolution in terms of a competition between energy and nearness of successive distributions in terms of a second coarse graining, resulting in a Fokker-Planck type equation.

ACKNOWLEDGEMENTS

The authors acknowldege the close collaboration of Richard James and Shlomo Ta'asan. This research was supported by the ARO and the NSF.

REFERENCES

[1]   Abeyaratne, R., Chu, C., and James, R.   1996, Phil. Mag. A, 73.2, 457-497
[2]   Jordan, R., Kinderlehrer, D., and Otto, F.   1997, Physica D, 107, 265-271
[3]   Jordan, R., Kinderlehrer, D., and Otto, F.   1998, SIAM J Math Anal, 29, 1-17
[4]   Jordan, R., Kinderlehrer, D., and Otto, F.   in preparation
[5]   Kinderlehrer, D.   1997 in Mathematics and Control in Smart Structures, (Varadan, V.K. and Chandra, J., eds) Proc. SPIE 3039, 2-13
[6]   Bensousson, A., Lions, J.-L., and Papinicolaou, G.   1978 Asymptotic analysis for periodic structures, North Holland
[7]   Ball, J.M. and James, R.D.   1987, Arch. Rat. Mech. Anal. 100, 13-52, 1991, Phil. Trans. Roy. Soc. Lond. A338, 389-450
[8]   James, R. and Kinderlehrer, D.   1989 in PDE's and continuum models of phase transitions, (Rascle, M., Serre, D., and Slemrod, M., eds.) Lecture Notes in Physics 344, Springer, 51-84.
[9]   Bhattacharya, K.   1992, Arch. Rat. Mech. Anal. 120, 201-244
[10]   Ericksen, J.L.   1987 in Metastability and Incompletely Posed Problems, (S. Antman, J.L. Ericksen, D. Kinderlehrer, I. Müller,eds) IMA Vol. Math. Appl. 3, Springer, 77-96
[11]   Finnis, M.W. and Sinclair, J.E. 1984 Phil. Mag. A 50, 45
[12]   Sutton, A.P. and Baluffi, R.W. 1995 Interfaces in Crystalline Materials, Oxford

THE INFLUENCE OF ANISOTROPIC GRAIN BOUNDARY ENERGY ON TRIPLE JUNCTION MORPHOLOGY AND GRAIN GROWTH

ALEXANDER H. KING

Department of Materials Science and Engineering, State University of New York at Stony Brook, Stony Brook, NY 11794-2275, U. S. A.

ABSTRACT

We consider some examples of triple junction equilibration in the presence of grain boundary energy anisotropy. It is shown that the presence of one or two cusp-trapped grain boundaries can reduce the restrictions upon the dihedral angles formed with the remaining (isotropic) boundaries This allows for a reduction in the average grain boundary curvature, and thus in the driving force for grain boundary migration.

INTRODUCTION

Triple junctions are the lines where three grain boundaries come together, and they are an essential feature of polycrystalline microstructure. Measurement of the dihedral angles at triple junctions has long been used as a means of determining the relative magnitudes of the grain boundary energies [1], although this has only been done using the assumption of isotropic grain boundary energy. Under this approximation, there is a unique solution for the dihedral angles if the energies of the three grain boundaries meeting at a triple junction are known; and maintaining the correct dihedral angles at all triple junctions in a polycrystal produces the interfacial curvature that drives grain boundary motion and, consequently, grain growth.

In this paper, we address the question of triple junction morphology for cases in which grain boundary energy anisotropy is significant. We have previously shown that there is a strong preference for symmetric boundary planes at triple junctions in thin-film gold [2] and one of our goals in this paper is to understand why this might occur. In attempting to understand this, we must consider the question of thermodynamic equilibrium at triple junctions, in the case where the anisotropy of interfacial energy cannot be ignored.

We have also shown experimentally that triple junctions can be sites of solute segregation, beyond that of the boundaries that they link [3], so it is likely that chemical pinning of the triple junctions can occur, and we consider means of including this, and other pinning forces, into the analysis. The effect of triple-junction "drag" has been considered in a phenomenological way by Galina *et al.* who show that a term for the junction mobility is readily developed and included in the expression for interface migration [4].

CONDITIONS FOR EQUILIBRIUM

We consider the case of three boundaries meeting at a straight triple junction. For the case of anisotropic grain boundary energy, and without considering any energy contributions from the triple junction itself, the equilibrium form of the triple junction is assumed to be given by solving Herring's formula [5]:

$$\sum_{i=1}^{3} \gamma_i \tau_i + \sum_{i=1}^{3} \frac{\partial \gamma_i}{\partial \phi_i} \tau_i \times n = 0 \qquad (1)$$

9

where γ_i is the interfacial tension of the boundary i, τ_i is a unit vector along the boundary at the triple junction and normal to it, n is a unit vector along the junction, and ϕ_i is an angle measuring the inclination of each boundary with respect to some reference line. The first term is basically equivalent to the (isotropic) force triangle relation while the second term expresses the grain boundary torque term.

King et al. [6] have shown that the dislocation content of a triple junction between three periodic grain boundaries can be a periodic function of the position of the junction, and this contributes a periodically varying elastic energy that may be associated with the junction, and treated as a component of the junction's excess energy. This locally varying energy behaves rather like the Peierls energy of a dislocation, and produces similar effects upon the triple junction. Yin et al. [3] have demonstrated that segregation occurs to triple junctions and it presumably affects the energy of the system if a triple junction is dragged away from its atmosphere of segregated solute, so this contributes an additional, non-periodic, position-dependent component of triple junction energy, which is of longer range than the lattice-dependent energy described above. This component has effects more like the pinning effect upon a dislocation of a solute atmosphere, and may produce breakaway effects akin to the solute drag effect upon a grain boundary [7] in which varying mobilities may be observed for a single triple junction.

We denote the total triple junction energy as W_t, per unit length, and any motion of the triple junction that results in changes of W_t is resisted or assisted by a force (per unit length of triple junction) given by

$$F = -\nabla W_t \qquad (2)$$

This results in a modification to Herring's formula, and the equilibrium form of the triple junction is now given by

$$\sum_{i=1}^{3} \gamma_i \tau_i + \sum_{i=1}^{3} \frac{\partial \gamma_i}{\partial \phi_i} \tau_i \times n - \nabla W_t = 0 \qquad (3)$$

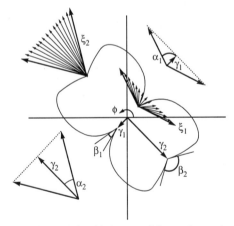

Figure 1: Schematic illustration of the relationship between ξ-fans and cusps in the Wulff plot. We use an idealized Wulff plot with two low-energy interfacial orientations. The deep, narrow cusp in the Wulff plot produces a short, wide ξ-fan, while the broad, shallow cusp produces a tall, narrow one. The fan is triangular, and for a symmetric boundary orientation it is an isosceles triangle with a half-angle, α, related to the angle between the Wulff-plot tangents at the cusp, by $\alpha = (\pi - \beta)/2$. The height of the fan triangle is γ.

Hoffman and Cahn have developed a vector thermodynamic treatment of surface energies [8,9] that can be used to form a graphical construction representing the equilibrium expressed in Eq. 1, or with a small modification, Eq. 3. The capillarity vector, ξ, corresponding to a given boundary plane orientation includes γ as its component normal to the boundary plane (just as in the Wulff construction) but also has a component lying in the boundary plane, which denotes the magnitude and principle direction of the torque, $\partial \gamma_i / \partial \phi_i$ that tends to rotate the boundary plane. The total capillarity vector is, in fact, perpendicular to the Wulff plot, at the corresponding interfacial orientation, and further details of the correspondence between ξ and the Wulff plot are given in Ref. 8. An advantage of the capillarity vector is that it contains information about all of the forces exerted upon the interface by its own energy.

It is important to note two particular features of ξ. *First*, the value of ξ may not be unique for the case of an orientation that produces a discontinuity in $\partial \gamma_i / \partial \phi_i$. When the boundary energy is a sharp minimum, with respect to inclination, ξ takes on a range of values defined by a constant value of γ, and a prescribed range of values of the torque term. As illustrated in Figure 1, the "fan" of ξ vectors is defined by the limiting normals to the Wulff plot at the energy cusp. Although the normal to the Wulff plot, which gives the orientation of ξ, might more correctly be considered as undefined at the cusp, for our purposes it is considered to include all of the values that would occur if the sharp change in slope were achieved by a continuous variation over an infinitesimal change of ϕ. If the energy gradients are steep at the cusp, then the ξ-fan is correspondingly broad, and if the energy gradients are shallow, describing a modest minimum in the interfacial energy, then the ξ-fan is narrow. The ξ-fan effectively describes the magnitudes of the applied torques that the interface will successfully resist. *Second*, ξ is properly described as a pseudo-vector rather than a true vector, since ξ and $-\xi$ are equivalent descriptors of the same interface, in the same state. This equivalence makes it necessary to adopt a particular convention for selecting the appropriate sign when considering equilibrium at triple junctions. We may assign an arbitrary line sense to the junction itself, then all of the vectors, ξ_i, should be drawn so they cross their corresponding interfaces in the same direction, *i.e.* that of a right-handed or a left-handed rotation about the triple junction line, as shown in Fig. 2.

Equilibrium forms of interfaces are found when all of the forces included in ξ are properly balanced. Graphically, this is achieved by drawing a closed vector-triangle in the case where the term in ∇W_t is negligible, corresponding to Eq. 1. When the forces imposed by the triple-junction self-energy are not negligible, as expressed in Eq. 3, then the vector triangle defined by ξ_i ($i = 1,2,3$)

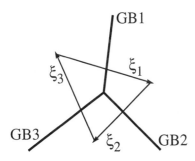

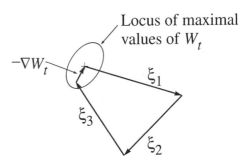

Figure 2: Graphical representation of the Herring formula (equation 1) using capillarity vectors. The three grain boundaries are labeled GB1, GB2 and GB3, and the respective capillarity vectors are drawn in a right-hand sense around the triple junction, forming a closed triangle.

Figure 3: The condition for equilibrium when the triple junction is affected by a pinning force. If the three capillarity vectors fail to form a closed triangle, the triple junction will still be stable provided that a large enough pinning force is available. The locus of maximum junction energy in each direction forms an envelope within which the closure failure of $\xi_1 + \xi_2 + \xi_3$ must fall.

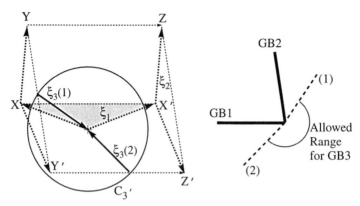

Figure 4. Allowed triple junction configurations for two cusp-trapped boundaries and one isotropic grain boundary. ξ_1 ends anywhere on the line XX', which forms the starting point for ξ_2. Since ξ_2 is also a fan of possible vectors, the allowed starting points for ξ_3 must fall in the parallelogram YY'Z'Z. With moderate energies for GB3 relative to those of GB1 and GB2, we deduce that $\xi_3(1)$ and $\xi_3(2)$ represent limiting configurations. The allowed inclinations of GB3 are illustrated in the sketch to the right.

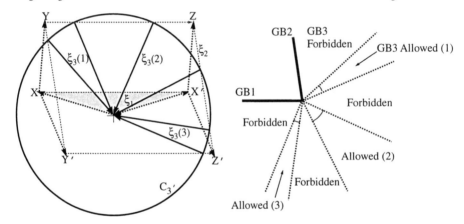

Figure 5. Allowed triple junction configurations for two cusp-trapped boundaries and one isotropic grain boundary, where the isotropic boundary has larger energy than for the case illustrated in Figure 4. Here we see that C_3' has as many as three distinct ranges of intersection with YY'Z'Z, corresponding to three allowed inclinations of GB3, as illustrated on the right.

needs only close within a region defined by the maximal magnitude of W_t, as shown in Fig. 3. Thus we see that the existence of a position-dependent self energy that resists triple junction motion can also stabilize configurations that would appear to be out of equilibrium, on the basis of the interfacial energy alone, by a margin that depends upon the magnitude of the stabilizing force.

APPLICATION TO PARTICULAR CASES

The case in which two out of three boundaries are trapped in energy-cusps is both interesting and instructive, and the reader will be able to extrapolate the analysis readily to other cases.

The appropriate ξ-triangle constructions are shown in Fig. 4. ξ_1 starts at the origin and can end anywhere on the line XX'. XX' then defines the locus of possible starting positions for ξ_2, which also takes the form of a fan. The possible ending points of ξ_2 then fall within the parallelogram YY'Z'Z, which defines the possible starting points of ξ_3. We use the circle C_3', centered at the origin and with radius equal to γ_3, to identify the possible values of ξ_3 that close the ξ-triangle (assuming no pinning force on the triple junction). The allowed values of ξ_3 are given by the range of intersection of C_3' with YY'Z'Z: when C_3' falls outside this area, no closed ξ-triangle can be formed. The result for relatively small moduli of ξ_3 is illustrated in Fig. 4. Boundary 3 can adopt any inclination to Boundaries 1 and 2, within an "allowed" range. There is no specific value of the dihedral angle that is characteristic of the energy of Boundary 3. If Boundary 3 moves outside the "allowed" range of inclinations, the result is a re-organization of the triple junction that pulls one or both of the anisotropic boundaries out of their low-energy facet orientations, unless there is a pinning force on the junction, able to resist the net forces applied by the interfaces themselves.

A more complicated case arises if we consider an isotropic grain boundary energy that is comparable to the ξ-fan half-widths associated with the cusp-trapped boundaries. A particular example is illustrated in Fig. 5, where it is shown that C_3' can have as many as three separate ranges of intersection with YY'Z'Z, generating three corresponding allowed ranges of inclination for Boundary 3 at the triple junction. Similar constructions demonstrating only two allowed ranges are quite easy to construct, and are left as an exercise for the reader.

When only one of the boundaries at a triple junction is cusp-trapped, the other two boundaries can adopt a range of dihedral angles, although they are related to each other in a fixed manner. In short, we find that the ability of cusp-trapped boundaries to sustain a range of applied torques results in a certain amount of freedom of inclination for other boundaries adjoined to them. If the triple junctions are pinned, as described above, then the extent of inclinational freedom is increased.

DISCUSSION - Effects on grain growth

We have previously reported that symmetric tilt grain boundaries are observed with unexpectedly high frequency in polycrystalline thin films of gold, annealed to the point of grain growth stagnation [2]. Approximately 70% of all grain boundaries are either "true" symmetric tilt boundaries or "symmetric-plus-double-positioning" boundaries where they meet triple junctions in these {111}-textured films. (The high incidence of symmetric-plus-double-positioning boundaries - comparable to that of symmetric tilt boundaries - leads us to presume that these pseudo-symmetric boundaries are energetically favored.) It is now possible to rationalize our observation, if it is assumed that these symmetric boundary orientations correspond to energy cusps of the types assumed in our foregoing analysis.

A high incidence of triple junctions incorporating cusp-trapped boundaries will have a profound and significant effect upon grain growth. The conventional view of grain growth may be summarized in the statement that it derives from curvature-driven grain boundary migration, and the grain boundary curvature derives from the preservation of particular dihedral angles at the triple junctions. The analysis presented here, shows that particular dihedral angles are not required when one or more of the boundaries is trapped in a particular orientation, and a range of inclinations of the isotropic boundary or boundaries may provide equally stable configurations. Under these conditions, boundary curvature may be removed completely, as shown in Fig. 6, and there is no driving force for the migration of the grain boundaries adjacent to these triple junctions. Thus, once such a triple junction is formed, it may be stable against removal by the normal processes of grain growth, unless a grain-switching event occurs at a neighboring triple junction. These junctions thus affect the process of grain growth in such a way that they are preserved in the microstructure, and they also contribute to grain growth stagnation as their concentration increases. Of course, the trapped interfaces themselves are also free of curvature, and do not contribute to grain growth, so

13

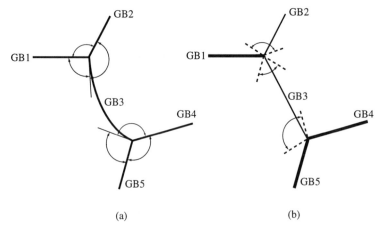

(a)   (b)

Figure 6: The effect of anisotropy on grain boundary curvature. When the dihedral angles at triple junctions are fixed, as in (a), then boundary curvature results for GB3 in this example. Using exactly the same locations for GB1, GB2, GB4 and GB5, but using allowed *ranges* of dihedral angle deriving from one or two of the boundaries being cusp-trapped, we find that GB3 can be free of curvature, as shown in (b). This removes any driving force for the migration of GB3, and thus will have a retarding effect on grain growth.

the most persistent triple junctions in the microstructure are likely to be those comprising three cusp-trapped boundaries.

CONCLUSIONS

When one or two of the interfaces at a triple junction are trapped in a particular orientation because of the shape of their respective energy surfaces, the remaining interfaces gain a degree of orientational freedom. This reduces interfacial curvature, in general, and lowers the driving force available for grain growth.

Acknowledgment: This work was supported by National Science Foundation grant number DMR 9530314.

REFERENCES

1.   C.G. Dunn and F. Lionetti, *Trans A.I.M.E.* **185**, 125 (1949).
2.   V. Singh and A.H. King, *Scripta Metall.* **34**, 1723-1727 (1996).
3.   K.M. Yin, A.H. King, T.E. Hsieh, F.R. Chen, J.J. Kai and L. Chang, *Microsc. and Microanalysis,* **3**, 417 (1997).
4.   A.V. Galina, V.E. Fradkov and L.S. Shvindlerman, *Fiz. Met. Metalloved.* **63**, 1220 (1987).
5.   C. Herring, in *The physics of powder metallurgy* (ed. W.E. Kingston) p.143 McGraw-Hill, New York (1951).
6.   A.H. King, F.R. Chen, L. Chang and J.J. Kai, *Interface Science* **5**, 287 (1997).
7.   J.W. Cahn, *Acta Metall.* **10**, 789 (1963).
8.   D.W. Hoffman and J.W. Cahn, *Surface Sci.* **31**, 368-388 (1972).
9.   J.W. Cahn and D.W. Hoffman, *Acta Metall.* **22**, 1205-1214 (1974).

LATTICE STATICS GREEN'S FUNCTION FOR MODELING OF DISLOCATIONS IN CRYSTALS

V.K. Tewary
Materials Reliability Division, National Institute of Standards and Technology, Boulder, CO 80303, vinod.tewary@nist.gov

ABSTRACT

A lattice statics Green's function method is described for modeling an edge dislocation in a crystal lattice. The edge dislocation is created by introducing a half plane of vacancies as in Volterra's construction. The defect space is decomposed into a part that has translation symmetry and a localized end space. The Dyson's equation for the defect Green's function is solved by using a defect space Fourier transform method for the translational part and matrix partitioning for the localized part. Preliminary results for a simple cubic model are presented.

INTRODUCTION

We describe a lattice statics Green's function (LSGF) method for modeling of dislocation in crystal lattices. The LSGF method was originally developed for point defects [1]. It has been applied [2,3] to small cracks consisting of 10-50 atoms but, so far, it has not been possible to use this method for large defects such as dislocations or cracks involving several hundred or more atoms [4]. In this paper, we describe a defect space Fourier transform method that enables us to apply the LSGF method to extended defects. This paper essentially describes our work in progress. Only the methodolgy and some preliminary results are reported here. Details will be published elsewhere.

The LSGF method is based upon the Kanzaki method and uses the Fourier representation of the perfect Green's function. Consequently it is possible to model a large crystallite containing several million atoms with a small CPU effort. Alternative methods for modeling dislocations use direct computer simulation based upon molecular dynamics [5, 6] or quasicontinuum [7]. Both these methods are very powerful and have important advantages. The advantage of the LSGF method is that it gives semi-analytical results for large crystallites and is useful for a quick determination of the basic physical characteristics of the defects. It is also useful for providing starting estimates for a detailed calculation using massive computer simulation of complicated defect structures.

BASIC FORMULATION

We consider a monatomic Bravais lattice. We assume a Cartesian frame of reference with an atomic site as origin. We denote the lattice sites by vector indices l, l', etc. The 3d force constant matrix between atoms at l and l' is denoted by $\phi^*(l, l')$. The force on atom l and its displacement from equilibrium position will be denoted, respectively, by $F(l)$ and $u(l)$, which are 3d column vectors.

Mat. Res. Soc. Symp. Proc. Vol. 529 © 1998 Materials Research Society

Following the method given in [1,2], we obtain

$$\mathbf{u(l)} = \Sigma \, \mathbf{G^*(l, l')} \, \mathbf{F(l')}, \tag{1}$$

where the defect Green's function

$$\mathbf{G^*} = [\phi^*]^{-1}. \tag{2}$$

The sum in eq. (1) is over all lattice sites and Cartesian coordinates, which has not been explicitly shown for brevity.

In the representation of the lattice sites, $\mathbf{G^*}$ and ϕ^* are $3N \times 3N$ matrices where N is the total number of lattice sites in the Born-von Karman supercell. For a perfect lattice in equilibrium without defects, $\mathbf{F(l)}$ is 0 for all l and the force constant and the Green's function matrices have translation symmetry. We denote these matrices by ϕ and $\mathbf{G}$ respectively. When a defect is introduced in the lattice, $\mathbf{F(l)}$ becomes, in general, nonzero and the force constant matrix changes. So

$$\phi^* = \phi - \Delta \, \phi, \tag{3}$$

where $\Delta\phi$ denotes the change in the ϕ. From eq. (3), we obtain the following Dyson equation

$$\mathbf{G^*} = \mathbf{G} + \mathbf{G} \, \Delta\phi \, \mathbf{G^*}, \tag{4}$$

where

$$\mathbf{G} = [\phi]^{-1} \tag{5}$$

is the perfect lattice Green's function.

For the perfect lattice, $\mathbf{G}$ is calculated by using the Fourier representation

$$\mathbf{G(l,l')} = (1/N) \, \Sigma_{\mathbf{q}} \, \mathbf{G(q)} \, \exp[i\mathbf{q}.(\mathbf{l\text{-}l'})], \tag{6}$$

where

$$\mathbf{G(q)} = [\phi(\mathbf{q})]^{-1}, \tag{7}$$

$\phi(\mathbf{q})$ is the Fourier transform of the force constant matrix and $\mathbf{q}$ is a vector in the reciprocal space of the lattice. For brevity of notations, we shall use the same symbol for a function and its Fourier transform, the distinguishing feature being the argument of the function. Since $\mathbf{G(q)}$ and $\phi(\mathbf{q})$ are 3×3 matrices, eqs. (9) and (10) can be used to calculate the $\mathbf{G(l,l')}$.

We define the defect space as the vector space generated by $\mathbf{l,l'}$ for which $\Delta\phi$ is nonvanishing. The lattice sites in the defect space will be denoted by λ,λ'. We partition

16

the matrices in eq. (4), and take only their components in the defect space. The Dyson equation in defect space is given by

$$g^* = g + g \, \Delta\phi \, g^*, \qquad (8)$$

where g, g^* are components of $\mathbf{G}$ and $\mathbf{G}^*$ in defect space. The matrices in eq. (8) are 3n x 3n matrices, where n is the number of atoms in the defect space. For point defects, n is small, so eq. (12) can be solved by direct matrix inversion. For extended defects such as dislocations and cracks, n is quite large. In earlier calculations, eq. (8) is solved numerically by restricting n to less than 50. In the computer simulation work, n is taken to be of the order 10-50, with N to be 50-100, which requires huge CPU effort. In the conventional LSGF method for cracks [3], although N can be 100-1000, n is typically less than 50. These values are unrealistically low. We show that by taking the Fourier transform in the defect space, we can solve eq. (8) for n=1000 or larger with minimal CPU effort.

Defect space Fourier transform method

We assume that the atoms interact through pair potentials and the interatomic potential is short range. This is a reasonable assumption since even the embedded atom potential that accounts for multibody interactions can be expressed as an effective dominant short range pair potential. For simple straight dislocations or planar cracks, the defect space has translation symmetry except near the ends. We exploit this translation symmetry in the defect space Fourier transform (DSFT) method for partially diagonalizing the Dyson equation.

Consider an edge dislocation. Although the DSFT method is applicable to 3d, for the sake of illustration in the present paper, we consider an infinite straight dislocation that makes the problem 2d. Following Volterra's construction, we create a half plane of vacancies, pull the atoms across the vacancy plane together by a distance u_c and bond them. Then we allow the lattice to relax to its new equilibrium position.

Figure 1 illustrates the model. Instead of a single edge dislocation, we create a dislocation dipole, so that the sum of all forces in the lattice is 0. A single dislocation implies unbalanced forces at the ends which makes the lattice unstable and introduces singularity in the displacement field. Atoms 1 - 4 are the created vacancies. The defect space includes all atoms in the box, 0-n, 0'-n', and 0''-n''. Notice that the defect space has translation symmetry except for the end atoms, 0,1, 4, n, and the corresponding primed and double primed atoms.

The displacement field for atoms in the defect space is given by

$$u(\lambda) = u_c + v(\lambda), \qquad (9)$$

where u_c is constant for all λ. We determine u_c by a minimization procedure and v by solving the Dyson equation. The determination of u_c includes nonlinear effects. We assume $v(\lambda)$ to be small and neglect cubic and higher order terms in v.

We write

$$\Delta\phi(\lambda,\lambda') = \Delta\phi_0(\lambda,\lambda') - \delta\phi(\lambda,\lambda'), \tag{10}$$

$$\mathbf{F}(\lambda) = \mathbf{F_0} + \mathbf{f}(\lambda), \tag{11}$$

where $\mathbf{F_0}$ and $\Delta\phi_0(\lambda,\lambda')$ are have translation symmetry. Hence, $\mathbf{F_0}$ is independent of λ and $\Delta\phi_0(\lambda,\lambda')$ depends upon λ, and λ' only through their difference. The end correction is given by $\mathbf{f}(\lambda)$ and $\delta\phi(\lambda,\lambda')$ which are nonvanishing only for atoms at the ends of the defect space. The atomic sites at the ends constitute the end space.

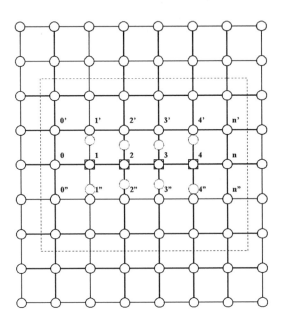

Fig. 1: Defect space for an edge dislocation dipole. Unrelaxed lattice sites are indicated by solid circles. Vacancies are created at sites marked by rectangles covering circles. The defect space consists of all sites inside the dashed rectangle. The end space in this model consists of sites 0,0',0", and n,n',n". The dotted circles show the relaxed positions of the atoms 1-4 calculated using the model given in [3]. The relaxation of other atoms is not shown.

We introduce Fourier transforms in the defect space as follows

$$\Delta\phi_0(\lambda,\lambda')=(1/n)\Sigma_k \Delta\phi_0(k) \exp[ik.(\lambda-\lambda')], \tag{12}$$

$$g(\lambda,\lambda')=(1/n) \Sigma_k g(k) \exp[ik.(\lambda-\lambda')], \tag{13}$$

where k takes n values between $-\pi$ and $+\pi$ such that $k.\lambda(n)$ is a multiple of π. The DSFT of the Green's function is given by

$$g(k) = \Sigma_\lambda g(0, \lambda) \exp(ik.\lambda), \tag{14}$$

or

$$g(k) = (1/N) \Sigma_q G(q) M(k,q), \tag{15}$$

where M is a projection function that projects the Green's function from the reciprocal space of the lattice to that of the defect space. It is given by

$$M(k,q) = \Sigma_\lambda \exp[i(k-q). \lambda]. \tag{16}$$

In general, the lattice sum in eq. (16) can be obtained analytically. It gives a discrete analogue of the Hilbert transform.

Using eq. (10) in eq. (8), and partitioning the matrix in the end space, we obtain for the Dyson equation in the end space

$$g^* = g_d - g_d \, \delta\phi \, g^* \tag{17}$$

where

$$g_d = [I - g \, \Delta\phi_0]^{-1} \, g. \tag{18}$$

We evaluate g_d using DFST given by eqs. (12) and (13) and then solve eq. (17) in the defect space by using matrix partitioning technique [1].

Results of a preliminary calculation are shown in Fig. 1. For these calculations, we assumed the same force constant model as given in [3]. Only the displacement field for atoms 1-4 in the defect space has been shown in Fig. 1. The displaced positions of these atoms are shown as dotted circles.

CONCLUSIONS

To summarize, the main advantage of the LSGF method using DSFT is that it is semianalytical, which allows us to model large crystallites and large defects with minimal CPU effort-- even for 3d dislocation problems. It can account for nonlinear effects locally in the defect space but assumes the harmonic approximation for atoms outside the defect space. Since the method involves independent sums over **q** and **k** space, the computational program can be easily vectorized if needed. The main disadvantages of our method are that it cannot give the time evolution of the equilibrium and it is limited to simple defect structures.

REFERENCES

1. V. K. Tewary, Green's function method for lattice statics, Adv. Phys. **22,** p757 (1973).
2. R. Thomson, S.J. Zhou, A.E. Carlsson, and V.K. Tewary, Lattice imperfections studied by use of lattice Green's functions, Phys. Rev. **B46**, p 10613 (1992).
3. V.K. Tewary and R. Thomson, Lattice statics of interfaces and interfacial cracks in bimaterial solids, J. Mater. Res. 7, p1018 (1992).
4. A.E. Carlsson and R. Thomson, Fracture toughness of materials: From atomistics to continuum theory, Solid State Physics **51**, p233 (1998).
5. P. Vashishta, A. Nakano, R.K. Kalia, et. al. Crack propagation and fracture in ceramic films- Million atom molecular dynamics simulations on parallel computers, Mat. Sci. Eng.- Solids **B37**, p56 (1996).
6. R. Pasianot, D. Farkas, and E.J. Savino, Dislocation core structure in ordered intermetallic alloys, J. Physique III **1**, p997 (1991).
7. E.B. Tadmor, M. Ortiz, and R. Phillips, Quasicontinuum analysis of defects in solids, Phil. Mag. **A73**, p1529 (1996).

THEORETICAL ANALYSIS OF FACETED VOID DYNAMICS IN METALLIC THIN FILMS UNDER ELECTROMIGRATION CONDITIONS

HENRY S. HO, M. RAUF GUNGOR, and DIMITRIOS MAROUDAS[a]
Department of Chemical Engineering, University of California, Santa Barbara, CA 93106-5080

ABSTRACT

A theoretical analysis is presented of the electromigration-induced dynamics of transgranular voids in metallic thin films. The analysis is based on self-consistent dynamical simulations of current-driven void surface propagation coupled with the distribution of the electric field in the metallic film. The simulation predictions highlight the rich nonlinear dynamics of current-driven evolution of voids that become faceted due to the strongly anisotropic nature of surface diffusion. The numerical results are analyzed based on approximate analytical solutions to faceted void migration and a linearized theory for the morphological stability of planar void facets.

INTRODUCTION

Microstructure evolution underlies a number of important materials reliability problems in microelectronics. Such a major problem is the electromigration-induced failure of polycrystalline aluminum and copper thin films, which are used for device interconnections in integrated circuits [1]. The continuous shrinking of the interconnect cross-sectional areas toward ultra-large-scale integration (ULSI) has led to conducting thin films with submicron-scale widths, which are characterized commonly by "bamboo" grain structure [1]. It has been established experimentally that transgranular voids are common sources of failure in bamboo films [1-5]; such voids are not associated with grain boundaries and nucleate usually at the film edges. Specifically, open-circuit failure occurs frequently due to slit-like features that emanate from transgranular voids and propagate fast across the film under the action of an electric field [1-5]. Recent theoretical analyses have addressed some aspects of current-driven transgranular void stability and dynamics [6-10].

The purpose of this paper is to analyze the electromigration-induced faceting, migration, and morphological stability and evolution of transgranular voids in metallic thin films. The analysis is based on a phenomenological formulation that captures the main physics of void dynamics and takes into account the strongly anisotropic nature of adatom surface diffusion in *fcc* metals. Numerical simulations highlight the extremely rich nonlinear dynamics of current-driven void evolution and its implications for thin-film reliability. The numerical results are analyzed based on approximate analytical solutions for the migration of stable faceted voids and linearized stability theory. Solutions that bifurcate from planar facet shapes also are discussed.

THEORETICAL ANALYSIS AND COMPUTATIONAL METHODS

Our analysis follows the continuum formalism developed in Refs. 8 and 9. Including both electromigration and curvature-driven surface diffusion gives the total surface atomic flux, J_s, as

$$J_s = \frac{D_s \delta_s}{\Omega \, k_B T} \, (-q_s^* E_s + \gamma \Omega \, \nabla_s \kappa). \tag{1}$$

In Eq. (1), D_s is the surface atomic diffusivity, Ω is the atomic volume, δ_s/Ω is the number of surface atoms per unit area, k_B is Boltzmann's constant, T is absolute temperature, q_s^* is a surface effective charge, E_s is the local component of the electric field tangential to the void surface, γ is the surface free energy per unit area, κ is the local surface curvature, and ∇_s is the surface gradient operator. The continuity equation gives the normal void propagation velocity, v_n, as

$$v_n = -\Omega \, \nabla_s \cdot J_s. \tag{2}$$

The analysis is limited in two dimensions, x and y; the void extends throughout the film thickness (in z), which is validated by experimental observations [5]. In Eq. (1), γ and q_s^* are assumed to be isotropic. This model describes most closely void dynamics in unpassivated films [5,8,9].

The electric field, $E(x,y)$, in the conductor is irrotational; thus, it can be written as $E=-\nabla\Phi$. The field also is solenoidal; thus, the potential Φ obeys Laplace's equation, $\nabla^2\Phi=0$. The void surface and the film's edges are modeled as insulating boundaries. A constant electric field, $E_\infty=E_\infty\hat{x}$, is imposed far away from the void surface. Dimensional analysis of Eqs. (1) and (2) gives two important dimensionless groups that govern void dynamics in the metallic film: the *surface electrotransport number* [9], $\Gamma\equiv(E_\infty q_s^*w^2)/(\gamma\Omega)$, that scales electric forces with capillarity forces, and the dimensionless void size, $\Lambda\equiv w_t/w$, that controls current crowding around the void surface. The width of the film, w, provides the characteristic length scale and w_t is the initial extent of the void across the film; the resulting time scale is $\tau=(k_BTw^4)/(D_s\delta_s\gamma\Omega)$. A value of $\Gamma=50$ corresponds to a current density of about 2 MA/cm^2 for $w=1$ μm and properties typical of pure aluminum; such current densities are typical of accelerated electromigration testing [3-5].

The anisotropy of surface adatom diffusivity is described by writing $D_s=D_{s,\min}f(\theta)$; the minimum surface diffusivity, $D_{s,\min}$, corresponds to a specific surface orientation, and $f(\theta)\geq1$ is a function of the angle θ formed by the local tangent to the surface and the direction of E_∞. We use $f(\theta)=1+A\cos^2[m(\theta+\phi)]$, where A, m, and ϕ are parameters that determine the strength of the anisotropy, the grain symmetry through the number of crystallographic directions that correspond to fast paths for surface diffusion, and the misorientation of a symmetry direction with respect to E_∞, respectively; $0\leq|\phi|\leq\pi/(2m)$. For the materials and the temperature range of interest, the anisotropy strength can reach values on the order of 10^4 [11].

In our numerical simulations, Laplace's equation is solved in the 2-D domain by a Galerkin boundary element method (BEM) employing linear trial functions; E_s is then derived from Φ for use in Eqs. (1) and (2). Eq. (2) is integrated in time explicitly to update the normal void surface displacement; a centered finite-difference scheme is used to calculate the surface flux divergence. Our BEM discretization is adaptive and employs several hundred nodes along the void surface. In our computations, if the void surface intersects the film's edge, the angle of intersection is set at 90°. This boundary condition is suggested by experimental observations [2-5]; this constraint does not affect our results qualitatively over a wide range of parameters [12].

RESULTS AND DISCUSSION

Our analysis of void morphological evolution at constant void volume is based on systematic search of the 5-D parameter hyperspace defined by the five dimensionless parameters Γ, Λ, A, m, and ϕ; this parametric search is the key in unraveling the most interesting nonlinear dynamical behavior. Our calculations have spanned the region of parameter space given by $0\leq\Gamma\leq500$, $0.1<\Lambda\leq0.8$, $1<A\leq10^4$, and $m=3, 2,$ and 1 for six-fold, four-fold, and two-fold grain symmetries, respectively; these are characteristic of <111>-, <100>-, and <110>-oriented grains, respectively.

Numerical Simulations of Electromigration-Induced Void Dynamics
Representative results of our numerical simulations are shown in Fig. 1. In cases of six-fold and four-fold symmetry, the dynamical behavior includes: formation of stable faceted voids that migrate at constant speed, facet selection mediated by elongation of two facets in expense of another one that they surround, and failure through propagation of fatal faceted slits across the film. If the void intersects the film's edge, facet selection leads to wedge-shaped voids that are either stable migrating at constant speed or extend and cause failure. Slit formation can occur easily in two-fold grain symmetry. Slit orientation depends strongly on both the anisotropy parameters and the experimental conditions. In general, high anisotropy strengths result in slit propagation across the film. In addition, voids may elongate significantly in the direction of the applied field. Furthermore, for the appropriate range of Γ, smaller voids can evolve into narrower slits. Generally, formation of slit-like features emanating from void surfaces is the outcome of an instability in the competition for mass transport between electromigration and capillarity.

Characterization of Void Mobility and Geometry
We have derived analytical expressions for the void migration speed as a function of the orientation of the void's facets, the void size, and the applied electric field. We focus on stable faceted voids that migrate at constant speed, $v_m\hat{x}$, and assume that the projection of the void morphology on the xy-plane can be described accurately as a closed polygon with N sides; note that in the absence of any facet selection prior to reaching steady state, $N=2m$ for the anisotropy

function under consideration. Let us consider the i-th facet of the void and denote its neighboring facets by i-1 and i+1, where the labeling increases in a clockwise sense; the corresponding orientations with respect to the horizontal far field direction, $\hat{x}$, are denoted by θ_i, θ_{i-1}, and θ_{i+1}, respectively. The speed of propagation normal to the i-th facet surface is $v_{n,i}=v_m\sin\theta_i$. For every facet, j, of the void, we approximate the tangential component of the electric field, $E_{s,j}$, by $E_{s,j}=\alpha_j E_\infty\cos\theta_j$, where α_j is a current crowding factor averaged over the length, l_j, of the j-th facet on the xy-plane. Integrating Eq. (2), over the length l_i gives

$$v_m \sin\theta_i \, l_i = \Omega \, (J_{s,i-1} - J_{s,i+1}). \qquad (3)$$

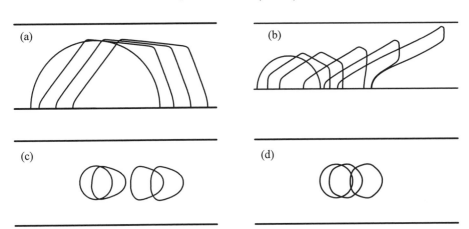

(a)

(b)

(c)

(d)

FIG. 1. Evolution of transgranular voids in a metallic film under the action of an electric field directed from left to right. The corresponding parameters and times for the different morphologies from left to right are: (a) $\Gamma=50$, $\Lambda=0.8$, $A=100$, $m=3$, $\phi=30°$; $t=0$, 0.486, 1.112, 1.739×10^{-4} τ; (b) $\Gamma=150$, $\Lambda=0.5$, $A=10$, $m=3$, $\phi=-15°$; $t=0$, 0.574, 1.272, 2.486, 3.395, 3.929×10^{-4} τ; (c) $\Gamma=50$, $\Lambda=0.4$, $A=10$, $m=2$, $\phi=45°$; $t=0$, 0.575, 2.527, 3.501×10^{-4} τ; (d) $\Gamma=50$, $\Lambda=0.4$, $A=10$, $m=3$, $\phi=0°$; $t=0$, 0.550, 1.769×10^{-4} τ. In (a) and (b), the void is initially semicircular and its surface intersects the edge of the film. In (c) and (d), the void is initially circular and is placed in the middle of the film.

For planar facets only electromigration contributes to the surface flux. The abrupt change in orientation at the void tips results in singularities, which cancel each other out and do not appear in Eq. (3). Using in Eq. (3) the local approximation for the electric field, $f(\theta)\neq 1$, and ignoring curvature-driven surface diffusion gives for each facet $i \in \{1,...,N\}$ that

$$v_m \sin\theta_i \, l_i = D_{s,\min}\delta_s q_s^* E_\infty \, [f(\theta_{i+1})\alpha_{i+1}\cos\theta_{i+1} - f(\theta_{i-1})\alpha_{i-1}\cos\theta_{i-1}]/k_B T. \qquad (4)$$

Eq. (4) provides the basis for characterizing the morphology and migration of stable faceted voids. Assuming that the facet lengths are known, Eq. (4) yields N nonlinear algebraic equations in N+1 unknowns: $\{\theta_i, i=1,...,N\}$ and the migrating speed, v_m. The constraint that the projection of the faceted void on the xy-plane is a closed polygon with N sides makes the approximate algebraic set well posed. Instead of solving the approximate equations, we manipulate Eq. (4) further to obtain characterization tools for our simulation results. A symmetric form for v_m can be obtained by adding Eqs. (4) for all the N facets of the void, which gives

$$v_m = \frac{D_{s,\min}\delta_s q_s^* E_\infty}{N k_B T} \sum_{i=1}^{N} \frac{f(\theta_{i+1})\alpha_{i+1}\cos\theta_{i+1} - f(\theta_{i-1})\alpha_{i-1}\cos\theta_{i-1}}{\sin\theta_i \, l_i}. \qquad (5)$$

Eq. (5) contains an important scaling result for characterization of stable faceted void migration, namely that $v_m \propto 1/l$, where l is a measure of the average facet length. Thus, Eq. (5) quantifies, for the case under consideration, the well known result that small voids migrate faster than large voids. Eq. (5) also implies that $v_m \propto (A_v)^{-1/2}$, where A_v is the area of the void in the xy-plane that is proportional to the void volume for the cylindrical voids considered in our 2-D modeling. We have tested the validity of this scaling relation based on our numerical simulations. Some representative results are shown in Fig. 2, showing that indeed the constant migrating speed of a stable faceted void scales with $(A_v)^{-1/2}$. The above results extend and generalize Ho's classical result for a circular void with isotropic surface properties migrating under electromigration conditions at constant speed inversely proportional to its radius, R [13]. Indeed, under isotropic conditions, $f(\theta)=1$, for the current crowding factor of a circular shape, $\alpha=2$ [9], and through integration of the continuity equation over a circular arc, as opposed to a planar facet, Eq. (5) reduces to the result obtained by Ho [13]: $v_m = 2D_s \delta_s q_s^* E_\infty / (k_B T R)$.

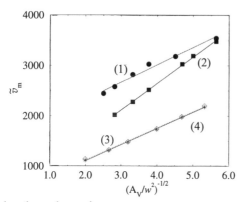

FIG. 2. Dependence of void migration speed on void size for stable faceted voids under electromigration conditions. The scaling relation $v_m \propto (A_v)^{-1/2}$ discussed in the text is verified for the cases: (1) $\Gamma=50$, $A=10$, $m=2$, $\phi=45°$; (2) $\Gamma=50$, $A=10$, $m=3$, $\phi=0°$; (3) $\Gamma=50$, $A=10$, $m=2$, $\phi=45°$; and (4) $\Gamma=50$, $A=10$, $m=3$, $\phi=-15°$. In (1) and (2), the void is initially circular and is placed in the middle of the film. In (3) and (4), the void is initially semicircular and is placed at the edge of the film.

lengths on the xy-plane:

Furthermore, adding all of Eqs. (4), for every facet $i \in \{1,...,N\}$, gives an expression for the faceted void morphological characteristics, relating facet orientations and

$$\sum_{i=1}^{N} \sin\theta_i \, l_i = 0. \tag{6}$$

Application of Eq. (6) to simulation results for dynamics of voids centered in the middle of the film's width yields summation values that fluctuate around zero with fluctuation amplitudes approximately equal to half a facet length; this suggests the validity of Eq. (6) for voids containing well-defined facets.

Analysis of Void Morphological Stability

The morphological stability of faceted voids is addressed in an inertial frame of reference moving at constant velocity, $v_m \hat{x}$, with respect to the laboratory frame (x,y,z), as predicted by Eq. (5): in this frame of reference, a stable void moving at constant speed v_m corresponds to a steady configuration. According to a Galilean transformation, the coordinates in the moving frame of reference are given by $x_m = x - v_m t$, $y_m = y$, and $z_m = z$. We consider a perturbation $h_m = y_m - m_i x_m$ from the planar facet morphology with orientation θ_i and slope $m_i = \tan\theta_i$. The corresponding transformation to express surface drift velocities, v_s, in the moving frame of reference is $v_{s,m} = v_s - v_m \cos\theta$. Hereafter, we are going to omit the extra subscript "m" for simplicity in the notation, while performing the analysis in the moving frame of reference.

Introducing the shape perturbation in the dimensionless form of the transformed continuity equation and linearizing about the steady morphology, $\theta = \theta_i$, $\tan\theta = m_i$ gives

$$\tilde{h}_{\tilde{t}} = -\hat{\alpha}_i \, \tilde{h}_{\tilde{x}\tilde{x}} - \beta_i \, \tilde{h}_{\tilde{x}\tilde{x}\tilde{x}\tilde{x}}. \tag{7}$$

In Eq. (7), scaled quantities, $\tilde{x} \equiv x/w$, $\tilde{y} \equiv y/w$, $\tilde{h} \equiv h/w$, and $\tilde{t} \equiv D_{s,min} \delta_s \gamma \Omega \, t/(k_B T w^4)$, are denoted by tildes, subscripts denote differentiation with respect to the corresponding variable, and the number of subscripts denotes the order of differentiation; $\hat{\alpha}_i$ and β_i are defined as

$$\hat{\alpha}_i \equiv (\Gamma \alpha_i + \tilde{v}_m) \, [m_i f(\theta_i) - df(\theta_i)/d\theta]/(1+m_i^2)^{3/2} \quad \text{and} \quad \beta_i \equiv f(\theta_i)/(1+m_i^2)^2, \tag{8}$$

24

where $\tilde{v}_m \equiv v_m k_B T w^2 /(D_{s,\,min} \gamma\Omega)$ is a dimensionless void migration speed. Taking the spatial Fourier transform of Eq. (7) gives the dispersion relation $\omega(k) = \hat{\alpha}_i k^2 - \beta_i k^4$, for the growth or decay rate of the shape perturbation $h(x,t) = u(x)\exp(\omega t)$. This relation and the definitions of Eq. (8) imply that a sufficient condition for the facet to be asymptotically stable, $\omega(k) < 0$, is

$$m_i f(\theta_i) - df(\theta_i)/ d\theta < 0. \qquad (9)$$

Setting $\omega(k_c) = 0$, gives the critical wavelength, λ_c, for the onset of instability as

$$\lambda_c \equiv 2\pi / k_c = 2\pi \{(\Gamma\alpha_i + \tilde{v}_m)(1 + m_i^2)^{1/2}[m_i - (df(\theta_i)/ d\theta)/ f(\theta_i)]\}^{-1/2}. \qquad (10)$$

We have characterized the morphological stability of faceted voids migrating at constant speed as predicted by our numerical simulations. In all cases, the sufficient condition of asymptotic stability, Eq. (9), was found to be valid for the planar facets of such voids. In addition, more detailed characterization based on Γ vs. m_i neutral stability curves according to Eq. (10) resulted in placement of the facets of such voids in the stable regime for planar morphologies.

Bifurcating Solutions from Wedge-Shaped Voids

We have also predicted surface waves that bifurcate from wedge-shaped void morphologies in low-dimensional regions of parameter space, where the planar morphology of the faceted shape becomes unstable. This is demonstrated in Fig. 3. The dynamical behavior of Figs. 3b (m=2) and 3d (m=1) can be computed by marching parallel to the Λ-axis in parameter space starting from the points that correspond to the states of Figs. 3a and 3c, respectively, and crossing over critical void sizes that define bifurcation points on these lines in 5-D hyperspace; the asymptotic states of Figs. 3a and 3c are wedge-shaped steady voids. The dynamics in Figs. 3b and 3d is characterized by the propagation on the surface of soliton-like features, which disappear gradually, while wedge-like failure occurs due to the significant current crowding effects at such void sizes in finite-width films. The details of the asymptotic nonlinear bifurcation analysis will be presented elsewhere. Current-induced appearance and disappearance of soliton-like features also has been predicted theoretically on an infinite initially flat free surface destabilized by an electric field [14].

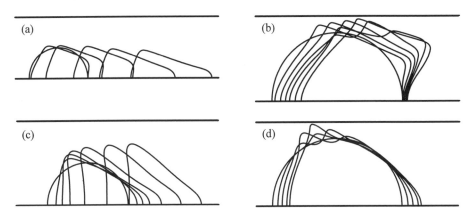

FIG. 3. Evolution of initially semicircular transgranular voids in metallic films with four-fold symmetry, m=2, (a) and (b), and two-fold symmetry, m=1, (c) and (d) under the action of an electric field directed from left to right. The corresponding parameters and times for the different morphologies from left to right are: (a) Γ=50, Λ=0.5, A=10, m=2, ϕ=45°; t=0, 0.021, 0.243, 0.699, 1.135, 1.645×10⁻⁴ τ; (b) Γ=50, Λ=0.8, A=10, m=2, ϕ=45°; t=0, 0.401, 0.813, 1.144, 1.401, 1.743, 2.147×10⁻⁴ τ; (c) Γ=50, Λ=0.5, A=10, m=1, ϕ=90°; t=0, 0.500, 0.997, 1.785, 3.575, 6.443, 9.160×10⁻⁴ τ; and (d) Γ=50, Λ=0.8, A=10, m=1, ϕ=90°; t=0, 0.381, 0.710, 1.093, 1.346×10⁻⁴ τ.

SUMMARY

In summary, we have presented a theoretical analysis of electromigration-induced void dynamics in metallic thin films based on self-consistent numerical simulations of void surface propagation coupled with the distribution of the electric field in the metal. Due to the strong dependence of surface diffusivity on surface orientation, void faceting is observed commonly; specifically, faceted voids that are stable and migrate at constant speed along the applied field direction are predicted. Our simulations also predict formation of wedge-shaped voids as the result of a facet selection mechanism, as well as the formation of slits that propagate fast across the film and cause failure. The migration of faceted voids at constant speed has been explained based on an approximate analytical theory, which gives a migration speed inversely proportional to a measure of the void size; this scaling relation is fully consistent with our simulation results. In addition, the morphological stability of the planar void facets has been analyzed based on linear stability theory implemented for a frame of reference moving at the constant void migration speed. The conclusions of the linear stability theory also are consistent with our numerical simulation results. Finally, surface waves traveling along void surfaces are predicted in certain regions of parameter space and have been discussed as solutions that bifurcate from planar surfaces of faceted voids. All of the above predictions are in excellent agreement with experimental observations [2-5].

ACKNOWLEDGMENTS

The authors acknowledge discussions with D. R. Clarke, S. J. Zhou, L. J. Gray, B. C. Larson, and S. T. Pantelides. This work was supported by NSF through a CAREER Award to DM (ECS-95-01111) and by the Frontiers of Materials Science Program of the UC Santa Barbara Materials Research Laboratory and Los Alamos National Laboratory (STB-UC:97-63).

REFERENCES

(a)   To whom correspondence should be addressed; E-mail: dimitris@calypso.ucsb.edu
1.   P. S. Ho and T. Kwok, Rep. Progr. Phys. **52**, 301 (1989); C. V. Thompson and J. R. Lloyd, Mater. Res. Soc. Bull. **18**, No. 12, 19 (1993).
2.   J. E. Sanchez, Jr., L. T. McKnelly, and J. W. Morris, Jr., J. Electron. Mater. **19**, 1213 (1990); J. Appl. Phys. **72**, 3201 (1992); J. H. Rose, Appl. Phys. Lett. **61**, 2170 (1992); Y.-C. Joo and C. V. Thompson (1997).
3.   O. Kraft, S. Bader, J. E. Sanchez, Jr., and E. Arzt, in Materials Reliability in Microelectronics III, edited by K. P. Rodbell, W. F. Filter, H. J. Frost, and P. S. Ho (Mater. Res. Soc. Symp. Proc. **309**, Pittsburgh, PA, 1993) pp. 199-204.
4.   E. Arzt, O. Kraft, W. D. Nix, and J. E. Sanchez, Jr., J. Appl. Phys. **76**, 1563 (1994).
5.   O. Kraft and E. Arzt, Acta Mater. **45**, 1599 (1997).
6.   Z. Suo, W. Wang, and M. Yang, Appl. Phys. Lett. **64**, 1944 (1994); W. Yang, W. Wang, and Z. Suo, J. Mech. Phys. Solids **42**, 897 (1994); W. Q. Wang, Z. Suo, and T.-H. Hao, J. Appl. Phys. **79**, 2394 (1996).
7.   O. Kraft and E. Arzt, Appl. Phys. Lett. **66**, 2063 (1995).
8.   D. Maroudas, Appl. Phys. Lett. **67**, 798 (1995).
9.   D. Maroudas, M. N. Enmark, C. M. Leibig, and S. T. Pantelides, J. Comp.-Aided Mater. Des. **2**, 231 (1995).
10.   L. Xia, A. F. Bower, Z. Suo, and C. F. Shih, J. Mech. Phys. Solids **45**, 1473 (1997).
11.   C.-L. Liu, J. M. Cohen, J. B. Adams, and A. F. Voter, Surf. Sci. **253**, 334 (1991).
12.   M. R. Gungor and D. Maroudas, to be published (1998).
13.   P. S. Ho, J. Appl. Phys. **41**, 64 (1970).
14.   M. Schimschak and J. Krug, Phys. Rev. Lett. **78**, 278 (1997).

LOCALIZED SURFACE INSTABILITIES
OF STRESSED SOLIDS

By J. COLIN, J. GRILHÉ, N. JUNQUA
Laboratoire de Métallurgie Physique, UMR 6630 CNRS, Université de Poitiers,
BP 179, F-86960 FUTUROSCOPE CEDEX, France

ABSTRACT

Localized instabilities formation on the free surface of solids has been studied when sources of non-homogeneous stress such as dislocations or precipitates are present in the bulk. This formalism of localized perturbations has been used to describe the butterfly transformation of cubic precipitates in superalloys and the contraction of rectangular specimens under stress.

INTRODUCTION

Roughness formation has been observed experimentally in many cases such as plate corrosion [1] or thin film growth on substrates [2-3]. At the same time, theoretical studies have demonstrated [4-12] that roughness development, by diffusion, on the free surfaces or interfaces of stressed solids can relax the stored elastic energy. One of the various methods proposed to characterize this roughness evolution is to develop the different profiles of surfaces in Fourier Series [5]. Considering the development of the total energy variation in powers of amplitude, the energy calculation has shown that [5-7], to the second order in amplitude, there is no interaction between the different harmonics. Thus, each harmonic of the Fourier series development of the surface profiles corresponding to sinusoidal instability, can be studied separately.

When a sinusoidal perturbation is introduced on the free surface of a solid submitted to a constant and homogeneous stress σ_0, the surface energy is increased of Δw^{surf} and the elastic energy is decreased of Δw^{elas} [6-7]. Above the critical wavelength λ_c defined by:

$$\lambda_c = \frac{2\pi\mu\gamma}{(1 - \nu)\,\sigma_0^2} \qquad (1)$$

where γ is the surface energy, ν the Poisson's ratio and μ the shear modulus of the material, the total energy variation is negative and sinusoidal instabilities of wavelength λ greater than the critical value λ_c can take place, by diffusion, on the surface of the stressed solid to relax the stored elastic energy: this is the Grinfeld instability.

Sinusoidal instabilities are well adapted to describe the beginning of the surface evolution under constant or homogeneous applied stress, but the study of a large class of problems such as linear defect nucleation, stressed plate striction or indentation involves non-homogeneous stress and needs a large number of harmonics to describe the localized surface fluctuation so that a localized perturbation method seems to be more adapted [13].

In this paper, the evolution of localized surface perturbations has been studied for different geometries of materials under different conditions of stress: semi-infinite solids, precipitates and rectangular plates [13,16].

Mat. Res. Soc. Symp. Proc. Vol. 529 © 1998 Materials Research Society

STUDY OF THE LOCALIZED INSTABILITIES EVOLUTION

A 2D planar surface is considered, see figure 1 for axes, with a planar deformation hypothesis ($\varepsilon_{zz} = 0$) and a constant stress tensor σ_0 with only a non-zero component $\sigma_{yy} = \sigma_0$ applied to the solid. A concentrated force $\vec{f} = (f_0,0,0)$ is also introduced at point $\vec{r_0} = (-c,0,0)$, with f_0 the force intensity per unit thickness.

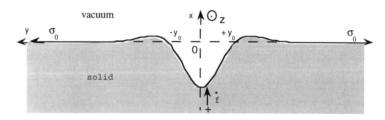

Figure 1. A wavelet instability has been introduced on the free surface of a stressed solid.

A fluctuation, spatially localized in the (Ox) direction, is supposed to appear by diffusion processes on the free surface and the corresponding free surface profile has been approximated to the so-called 2D *Mexican hat*, i.e the second derivative of the gaussian function [14]:

$$h(y) = a \left[\frac{y^2}{y_0^2} - 1 \right] \exp \left[-\frac{1}{2} \frac{y^2}{y_0^2} \right] \tag{2}$$

with a the amplitude and y_0 the opening parameter of the wavelet playing the role of the wavelength of sinusoidal instabilities. This symmetrical fluctuation of the surface which induces no variation of volume, seems to be well adapted to describe surface evolution under non-homogeneous stress [13].

When such a perturbation is introduced, the equilibrium condition of forces on the free surface is no longer fulfilled. In order to satisfy this boundary condition, a relaxation stress Σ_{rel} needs to be added:

$$\left(\Sigma_0^1 + \Sigma_0^2 + \Sigma_{rel} \right) \vec{n} = 0 \tag{3}$$

with $\vec{n}$ the normal vector to the free surface, $\vec{n} = (1,-h'(y),0)$.

The determination of Σ_{rel} can be performed by considering two distributions of Green functions or edge dislocations on the free surface [15]. In this paper, two distributions of edge dislocations of Burgers vectors respectively (b,0,0) and (0,b,0) have been used and the relaxation stresses have been calculated by developing equation (3) to the first order of the amplitude a. Calculating all energies per unit thickness to a second order in a, the total energy variation can be written as:

$$\Delta w^{\text{tot}} = \frac{\gamma a^2}{c} \left[\frac{15}{16} \sqrt{\pi} \frac{1}{\eta} - \frac{f_0^2}{\pi \mu \gamma (1-\nu)} \frac{1}{\eta^4} \psi(\eta) - \sqrt{\frac{2}{\pi}} \frac{\sigma_0 f_0}{\mu \gamma} \frac{1}{\eta^2} \chi(\eta) - \frac{(1-\nu) \sigma_0^2 c}{\mu \gamma} \right] \tag{4}$$

where μ is the elastic modulus, ν Poisson's ratio, $\chi(\eta)$ and $\psi(\eta)$ are numerically determined functions [13] and $\eta = y_0 / c$ is a dimensionless parameter.

The study of the total energy variation Δw^{tot} shows [13] that in a stressed solid free of defects $\left(f_0 = 0, \sigma_0 \neq 0 \right)$, the free surface is unstable in front of the fluctuations for parameters:

$$y_0 \geq y_0^c = \frac{16}{15} \sqrt{\pi} \, \frac{\mu\gamma}{(1 - \nu) \sigma_0^2} = \frac{8}{15\sqrt{\pi}} \lambda_c \qquad (5)$$

where y_0^c is of the same order as the critical wavelength λ_c of sinusoidal fluctuations. The maximum gain in energy is then obtained for large values of y_0. When the concentrated force only is considered ($\sigma_0 = 0$), a minimum in energy appears at $y_0^m \approx 5$ c for $|f_0^c| = 2.02 \sqrt{\mu\gamma c (1 - \nu)}$. The energy variation corresponding to this minimum becomes negative when $|f_0| \geq |f_0^p| = 2.32 \sqrt{\mu\gamma c (1 - \nu)}$. Under these conditions, the free surface of the solid can develop a ``mexican hat instability'' with the most probable parameter $y_0^c \leq 5$ c corresponding to the minimum in energy.

SUBDIVISION OF PRECIPITATES IN SUPERALLOYS

A 2D square precipitate in an infinite matrix is considered (figure 2). In this study, each 2h length interface of the precipitate is considered separately and the different interface interactions due to the geometry are neglected [16]. The lattice parameters, the shear modulus, Poisson's ratio and the biaxial modulus are respectively a_i, μ_i, ν_i and $E_i = 2\mu_i \frac{1 + \nu_i}{1 - \nu_i}$ in each material.

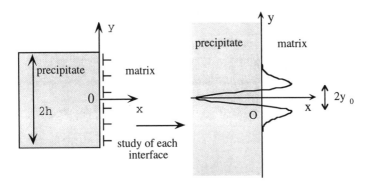

Figure 2. A 2D square precipitate is considered in an infinite matrix. Each interface of the precipitate has been studied separately.

Epitaxy between two materials of different lattice parameters and of different elastic constants, joined at a coherent interface, induces stresses in the neighbourhood of this interface of length 2h. According to previous studies [2-3], it can be anticipated that the epitaxial stress step, $\delta\sigma_{yy} = \frac{2}{1 + \nu} \frac{E_m E_p}{E_m + E_p} \frac{\delta a}{a} = 2\sigma_0$ through the interface, is the origin of the elastic relaxation of the perturbed interface. The relaxation stress induced by the wavelet shape-like deformed interface (figure 1) has been determined expressing the equilibrium of forces and the continuity of displacements at the interface [16]. The total energy variation is then found to be:

$$\Delta w^{tot} = \frac{\gamma e^2}{h} \left[\frac{15}{16} \sqrt{\pi} \, \frac{h}{y_0} + K \, \zeta(y_0 / h) \right] \qquad (7)$$

with the dimensionless constants,

$$K = \frac{1}{\pi} \frac{\sigma_0^2 h}{\alpha \gamma}, \quad \alpha = \frac{(\mu_p \kappa_m + \mu_m)(\mu_m \kappa_p + \mu_p)}{\mu_m \kappa_p (1 + \kappa_m) + \mu_p \kappa_m (1 + \kappa_p)} \tag{8}$$

and $\zeta(y_0 / h)$ a function which has been determined numerically. The study of the Δw^{tot} variation with y_0 / h [12] shows that the interface fluctuation is completely localized on a 2h interface when the opening parameter y_0 satisfies the relation $6y_0 = 2h$. When the interface length 2h is greater than the critical value defined by $2h_c = 3.16 \, \alpha \, \dfrac{\mu \gamma}{\sigma_0^2}$, the total energy variation is negative and the instability is completely localized on the interface. A good approximation of the critical size can be taken as $2h_c$ under which the precipitates of superalloys are stable (figure 3).

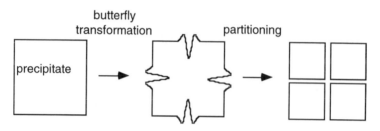

Figure 3. Butterfly transformation and partitioning of precipitates in a nickel-based superalloy.

Finally, it can be concluded that, when the length of each stressed precipitate-matrix interface is larger than the critical length $2h_c$, the initial 2D square precipitate (figure 3) can be transformed into a typical butterfly distorted cube and then partitioned [17-19]. The cube-octet transformation of a three-dimensional cubic precipitate can be deduced from this two-dimensional linear study.

CONTRACTION OF A RECTANGULAR PLATE UNDER STRESS

A 2D rectangular plate submitted to a constant stress σ_0 is considered (figure 4).

Figure 4. Contraction of a rectangular plate due to localized instabilities.

The shear modulus and the Poisson's ratio of the 2h width plate are μ and ν respectively. It is supposed that symmetrical localized wavelet instabilities can appear by diffusion on the two lateral free surfaces S_1 and S_2 of the plate (figure 4). The elastic energy variation induced by these perturbations can be calculated, to the second order in amplitude a, using two distributions of surface dislocations. The calculation of the total energy variation gives:

$$\Delta w^{tot} = \frac{\gamma e^2}{h} \left[\frac{30}{16} \sqrt{\pi} \frac{1}{\xi} - \Gamma \, \theta(\xi) \right]$$

with $\xi = y_0 / h$, $\Gamma = \dfrac{2(1 - \nu) \sigma_0}{\mu \gamma}$ and $\theta(\xi)$ a numerically determined function.

The study of the energy variation with y_0 / h shows that when the applied stress σ_0 is small, $\sigma_0 \ll \sqrt{\mu\gamma / h}$, the energy variation is always positive. In this case, the applied stress σ_0 is not sufficient for contraction to develop. For a particular value of the stress, $\sigma_0^p \approx \sqrt{\mu\gamma / h}$, a minimum in energy appears at $y_0^p \approx 2.5$ h and for a critical stress $\sigma_0^c \approx \sqrt{1.08\, \mu\gamma / h}$ slightly superior to σ_0^p, the energy variation corresponding to this minimum $y_0 = y_0^p$ is zero. Above this critical stress σ_0^c, each of the two free surfaces will develop wavelet instabilities of opening parameter $y_0^p < 2.5$ h which minimizes the energy.

This linear study of localized instabilities development, which only charaterizes the beginning of the surfaces evolution, has demonstrated the possibility for contraction of the plate to appear when the applied stress σ_0 is greater than the critical stress σ_0^c.

CONCLUSIONS

The formation of localized instabilities has been studied on the free surface of a stressed solid. An energy variation calculation has demonstrated that the development of these fluctuations is energetically favourable when solids are submitted to a non-homogeneous stress. The subdivision of 2D square precipitates in superalloys has been studied; a critical interface length of the precipitates $2h_c$ has been determined. Above this critical length $2h_c$, the precipitates in a matrix can develop localized instabilities on each of their interfaces and then partition. The contraction of two dimensional stressed plates has been finally characterized. The energy variation calculation has shown that, above a critical stress σ_0^c, the development of two symmetrical localized instabilities on the lateral free surfaces of a stressed rectangular plate is favourable and contraction can occur with an extension $y_0 < 2.5$ h.

ACKNOWLEDGEMENTS

The authors would like to thank L. CARTZ for reading the manuscript and making useful suggestions.

REFERENCES

1. R. J. Asaro and W. A. Tiller, Metall. Trans. **3**, pp. 1789-1796 (1972).

2. F. K. LeGoues, M. Copel and R. M. Tromp, Phys. Rev. B **42**, p. 11690 (1990).

3. S. Guha, A. Madhukar and K. C. Rajkumar, Appl. Phys. Lett. **57**, p. 2110 (1990).

4. M. A. Grinfeld, Dokl. Akad. Nauk. **290**, p. 1358 (1986).

5. P. Nozières, J. Physique I **3**, p. 1 (1993).

6. D. J. Srolovitz, Acta Metall. **37**, p. 621 (1991).

7. J. Grilhé, Acta Metall. **41**, p. 909 (1993).

8. C. H. Chiu and H. Gao, Int. J. Solids Struct. **21**, p. 2983 (1993).

9. B. J. Spencer, P. W. Voorhees and S. H. Davis, J. Appl. Phys. **73**, p. 955 (1993).

10. F. Jonsdottir, J. Modell. Simul. Mater. Sci. Eng. **3**, p. 503 (1995).

11. H. Gao, J. Mech. Phys. Solids **42**, p. 741 (1994).

12. J. Colin, N. Junqua and J. Grilhé, Acta Mater. **9**, p. 3855 (1997).

13. J. Colin, N. Junqua and J. Grilhé, Europhys. Lett. **38**, p. 307 (1997).

14. A. Grossmann and J. Morley, SIAM J. Math. Analysis **15**, p. 723 (1984).

15. K. Jagannadham and M. J. Marcinkowski, Physica Status Solidi (a) **50**, p. 293 (1978).

16. J. Colin, N. Junqua, J. Grilhé, Acta Mater. **46**, pp. 1249-1255 (1998).

17. J. F. Ganghoffer, A. Hazotte, S. Denis and A. Simon, Scripta Metall. **25**, p. 2491 (1991).

18. A. G. Khachaturyan, S. V. Semeovskaya and J. W. Morris, Acta Metall. **36**, p. 1563 (1988).

19. Y. Wang, L. Q. Chen and A. G. Khachaturyan, Scripta Metall. **25**, p. 1387 (1991).

ON THE INTERACTION OF INSOLUBLE SPHERICAL PARTICLES WITH A SOLIDIFYING INTERFACE IN THE PRESENCE OF MORPHOLOGICAL INSTABILITIES

L. HADJI* and A.M.J. DAVIS;
Mathematics Department, University of Alabama,
Tuscaloosa, AL 35487-0350, * lhadji @ UA1VM.UA.EDU

ABSTRACT

The interaction of a dilute suspension of spherical particles with a moving solidifying front is investigated analytically. The front emerges during the unidirectional solidification of a dilute binary alloy. A linear stability analysis of the planar interface reveals that the presence of the particles has a destabilizing effect, i.e. the critical value of the control parameter, taken here to be the concentration faraway from the interface, for the onset of the Mullins-Sekerka instability is lowered due to the particles' presence.

INTRODUCTION

The interaction of spherical particles with a solid-liquid interface arises in several physical and industrial situations. This is the case, for example, of the freezing of water in the presence of soil particles during the formation of frost heave [1]–[2], the cryopreservation of biological cell suspensions [3] and in casting wherein binary substances are doped with a small volume fraction of second phase inclusions during the solidification processing of Metal Matrix Composites. This process will yield a final cast product with enhanced material properties provided the added inclusions end up distributed uniformly in the solidified material. The latter task is difficult to accomplish due to the migration of the particles in the melt during the processing stage. Experiments have shown that the distribution of the particles in the cast depends primarily on the interaction of the particles with the advancing solid-liquid interface. Two main scenarios are found to possibly occur: either the particles are pushed or they are egulfed by the moving front. Several one-particle models have been analyzed which suggest the existence of a critical growth velocity above which the particles are engulfed and below which they are pushed [4]–[6]. These studies have also investigated the dependence of this critical velocity on the various physical and processing parameters.

The aim of this paper is to investigate the effect of the presence of the particles on the threshold parameters for the onset of interfacial instabilities, namely we determine the dependence of the critical value of the control parameter (the solute concentration faraway from the interface) and the corresponding critical wavelength on the volume fraction of particles in the mixture.

THEORY

Mathematical model

Consider the directional solidification of a dilute binary alloy in such a way that the solid-liquid interface is moving vertically upward with velocity V. The system is completely described by the heat conduction equations in the solid and liquid layers and the solute diffusion equation in the liquid. Heat diffusion in both phases is assumed to be quasi-instantaneous compared with solute diffusion, measured by the solute diffusion coefficient D, in the liquid phase. Lengths, time, temperature and concentration are scaled by D/V, D/V^2, T_m and C_∞, respectively, where T_m is the freezing temperature of the pure substance and C_∞ is the concentration far ahead of the front. In a coordinate system that has z measured into the liquid from the moving interface, the following system is obtained when the quasi-steady state assumption is evoked.

$$\nabla^2 T_L = \nabla^2 T_S = 0, \tag{1}$$

$$\nabla^2 C + \frac{\partial C}{\partial z} = 0. \tag{2}$$

At the interface $z = \eta(x, y, t)$, the continuity of temperature and the coupling between the temperature and concentration fields, which accounts for the Gibbs-Thompson effect, imply

$$T_S = T_L = T_M + mC - \gamma \mathcal{K}. \tag{3}$$

The conservation of heat at the interface leads to a balance between the jump in the heat flux and latent heat release

$$\frac{\partial}{\partial n} \left[\frac{K_S}{K_L} T_S - T_L \right] = \mathcal{L} \left(1 + \frac{\partial \eta}{\partial t} \right). \tag{4}$$

The conservation of solute at the interface, with the added assumption of no solute diffusion in the solid phase, yields

$$\frac{\partial C}{\partial n} = (k - 1)(1 + \frac{\partial \eta}{\partial t})C. \tag{5}$$

Faraway from the interface:

$$\text{as } z \to \infty, \quad \frac{\partial T}{\partial z} = G_L, \quad C = 1, \quad \text{and} \quad \text{as } z \to -\infty, \quad \frac{\partial T_S}{\partial z} = G_S, \tag{6}$$

and local thermodynamic equilibrium at the interface requires $C_s = kC$. Here K_S and K_L are the thermal conductivities of the solid and liquid phases respectively, k is the partition coefficient, $\gamma = \Gamma V/D$ is the surface tension, $\mathcal{L}$ is the dimensionless latent heat of fusion, $\mathcal{K}$ is the curvature of the interface, C_s is the concentration of solute in the solid phase and $\partial/\partial n$ is the normal derivative.

There exist basic solutions with a planar interface $\eta = 0$. The temperature and concentrations fields are diffusive and have profiles that depend on the vertical cooordinate only and satisfy the relations

$$T_{L0}(z) = G_L z + 1 + M/k, \tag{7}$$
$$T_{S0}(z) = G_S z + 1 + M/k, \tag{8}$$
$$C_0(z) = 1 - \frac{k-1}{k} e^{-z}. \tag{9}$$

where $M = mc_\infty/T_m$ is the dimensionless liquidus slope.

Effective distribution coefficient

In the absence of particles, the ambient concentration C_0 in the fluid region $z > 0$, where $z = 0$ is the moving interface is given by Eq. (9). The insertion of particles into the fluid will necessitate modifications to the concentration C from its ambient value $C_0(z)$ due to the requirement that C have zero normal derivative on each particle surface. It will be assumed that the particle volume fraction ϕ is small enough for the $O(\phi)$ effects to suffice.

Suppose first that a single particle of radius a is introduced and centered at $(x, y, z) = (0, 0, z_0)$, where $z_0 > 0$. Let local spherical coordinates (r, σ) be defined by

$$r = \sqrt{x^2 + y^2 + (z - z_0)^2}, \qquad z - z_0 = r\sigma \tag{10}$$

and, ignoring for the present the interface condition Eq. (5), seek to construct a suitably accurate disturbance concentration C_1 that satisfies the normal derivative condition

$$\left(\frac{\partial C_1}{\partial r}\right)_{r=a} = -\left(\frac{dC_0}{dz}\right)_{z=z_0} \sigma e^{-a\sigma}. \tag{11}$$

Evidently, C_1 will be singular at $r = a$ and have scale factor a^3. Separated solutions of the form

$$C_1 = e^{-\frac{1}{2}(z+r)} q_n(r) P_n(\sigma), \tag{12}$$

where P_n denotes a Legendre polynomial, exist provided q_n satisfies

$$\frac{d^2 q_n}{dr^2} - \left(1 - \frac{2}{r}\right)\frac{dq_n}{dr} - \left[\frac{1}{r} + \frac{n(n+1)}{r^2}\right] q_n = 0, \tag{13}$$

whence rq_n must be a polynomial of degree n in r^{-1}, namely,

$$rq_n = \sum_{s=0}^{n} \frac{1}{r^s} \frac{(n+s)!}{s!(n-s)!}. \tag{14}$$

In particular,

$$rq_0 = 1, \qquad rq_1 = 1 + \frac{2}{r}, \qquad rq_2 = 1 + \frac{6}{r} + \frac{12}{r^2}. \tag{15}$$

Note that the highest powers of r^{-1} dominate at small r, i.e.

$$q_n \sim \frac{1}{r^{n+1}} \frac{(2n)!}{n!} \quad \text{as } r \to 0. \tag{16}$$

Hence the perturbation concentration C_1 can be written as

$$C_1 = \left(\frac{dC_0}{dz}\right)_{z=z_0} e^{-\frac{1}{2}(z-z_0+r)} \sum_{n=0}^{\infty} A_n a^{n+2} q_n(r) P_n(\sigma), \tag{17}$$

from which the normal derivative condition yields

$$\sigma e^{\frac{1}{2}a(1-\sigma)} = \sum_{n=0}^{\infty} A_n a^{n+2} \left[\frac{1+\sigma}{2} q_n(a) - q_n'(a)\right] P_n(\sigma). \tag{18}$$

The orthogonality of $\{P_n(\sigma); n \geq 0\}$ now enables the formula for C_1 to be reduced to

$$C_1 = \left(\frac{dC_0}{dz}\right)_{z=z_0} e^{-\frac{1}{2}(z-z_0+r)}\frac{a^3}{4r}\left[\left(1+\frac{2}{r}\right)\sigma - 1\right] + O(a^5),$$

$$= \left(\frac{dC_0}{dz}\right)_{z=z_0}\frac{a^3}{2}\left(1 - \frac{\partial}{\partial z_0}\right)\left[-\frac{e^{-\frac{1}{2}(z-z_0+r)}}{r}\right] + O(a^5). \tag{19}$$

This form of the singularity is very helpful in considering a periodic array of such singularities because the Fourier transform of $r^{-1}e^{-\frac{1}{2}(z-z_0+r)}$ is readily available, namely

$$\frac{1}{r}e^{-\frac{1}{2}(z-z_0+r)} = \frac{e^{-\frac{1}{2}(z-z_0)}}{2\pi}\int_{-\infty}^{\infty}\int_{-\infty}^{\infty}\frac{e^{-(\nu^2+\frac{1}{4})^{1/2}|z-z_0|}}{(\nu^2+\frac{1}{4})^{1/2}}e^{i(\nu_1 x+\nu_2 y)}d\nu_1 d\nu_2, \tag{20}$$

in which $\nu^2 = \nu_1^2 + \nu_2^2$. Hence

$$C_1 = \left(\frac{dC_0}{dz}\right)_{z=z_0}\frac{a^3}{4\pi}e^{-\frac{1}{2}(z-z_0)}\int_{-\infty}^{\infty}\int_{-\infty}^{\infty}e^{-(\nu^2+\frac{1}{4})^{1/2}|z-z_0|}$$

$$\left[\frac{|z-z_0|}{z-z_0} - \frac{1}{(4\nu^2+1)^{1/2}}\right]e^{i(\nu_1 x+\nu_2 y)}d\nu_1 d\nu_2 + O(a^5). \tag{21}$$

This equation gives the perturbed concentration C_1 in a form that readily facilitates the extensions required here. First, the imposition of the interface condition yields an image singularity, viz

$$C_1^* = \left(\frac{dC_0}{dz}\right)_{z=z_0}\frac{a^3}{4\pi}e^{-\frac{1}{2}(z-z_0)}\int_{-\infty}^{\infty}\int_{-\infty}^{\infty}\left\{e^{-(\nu^2+\frac{1}{4})^{1/2}|z-z_0|}\left[\frac{|z-z_0|}{z-z_0} - \frac{1}{(4\nu^2+1)^{1/2}}\right]\right.$$

$$\left. -e^{-(\nu^2+\frac{1}{4})^{1/2}(z+z_0)}\left[1 + \frac{1}{(4\nu^2+1)^{1/2}}\right]\frac{(4\nu^2+1)^{1/2}-2k+1}{(4\nu^2+1)^{1/2}+2k-1}\right\}e^{i(\nu_1 x+\nu_2 y)}d\nu_1 d\nu_2 + O(a^5). \tag{22}$$

Next, the corresponding perturbed concentration $C_\infty^*(z_0)$ due to a doubly periodic array of such singularities at $(2nL, 2mL, z_0)$ $(-\infty < n, m < \infty)$ is deduced by using the identity

$$\sum_{-\infty}^{\infty}\sum_{-\infty}^{\infty}e^{i\nu_1(x-2nL)}e^{i\nu_2(y-2mL)} = \frac{\pi^2}{L^2}e^{i(\nu_1 x+\nu_2 y)}\sum_{-\infty}^{\infty}\sum_{-\infty}^{\infty}\delta\left(\nu_1 - \frac{n\pi}{L}\right)\delta\left(\nu_2 - \frac{m\pi}{L}\right) \tag{23}$$

to obtain

$$C_\infty^*(z_0) = \left(\frac{dC_0}{dz}\right)_{z=z_0}\frac{a^3\pi}{4L^2}e^{-\frac{1}{2}(z-z_0)}\sum_{-\infty}^{\infty}\sum_{-\infty}^{\infty}\left\{e^{-\frac{1}{2}\kappa_{nm}|z-z_0|}\left[\frac{|z-z_0|}{z-z_0} - \frac{1}{\kappa_{nm}}\right]\right.$$

$$\left. -e^{-\frac{1}{2}\kappa_{nm}(z+z_0)}\left[1 + \frac{1}{\kappa_{nm}}\right]\frac{\kappa_{nm}-2k+1}{\kappa_{nm}+2k-1}\right\}e^{\frac{i\pi}{L}(nx+my)} + O(a^5), \tag{24}$$

where

$$\kappa_{nm}^2 = 1 + \frac{4\pi^2}{L^2}(n^2 + m^2). \tag{25}$$

Finally, the total concentration C in the liquid region exterior to a periodic cubic array of spheres on which the non-penetration condition is applied, is given by

$$C = C_0 + \sum_{j=1}^{\infty}C_\infty^*(2jL). \tag{26}$$

The mean concentration $\bar{C}(z)$ in a plane parallel to the interface is therefore given by

$$\bar{C} = C_0 - \frac{a^3\pi}{2L^2} \sum_{j=1}^{\infty} \left(\frac{dC_0}{dz}\right)_{z=2jL} \left[H(z - 2jL) - \frac{k-1}{k}e^{-z}\right] + O(a^5), \tag{27}$$

where H denotes the Heaviside unit function. On averaging over intervals of length $2L \ll 1$, it follows that the average concentration is

$$\langle C \rangle \sim 1 - \frac{k-1}{k}e^{-z}\left[1 + \frac{a^3\pi}{4L^3k}\right], \tag{28}$$

which yields, provided $\phi/k \ll 1$, the effective segregation coefficient

$$k_{eff} \sim k\left[1 + \frac{k-1}{k}\frac{a^3\pi}{4L^3}\right] = k + \frac{3}{2}(k-1)\phi. \tag{29}$$

RESULTS

The threshold value for the onset of instability of the planar solidifying front will then be altered due to the change in the segregation coefficient. For simplicity, we consider the condition of constitutional supercooling [7]

$$V = G/[-m(C_\infty/Dk)(1 - k)]. \tag{30}$$

When k in Eq. (30) is replaced by k_{eff}, and for $(\phi/k) \ll 1$ and $k = O(1)$, the instability threshold becomes

$$V = G/\{-m(C_\infty/Dk)(1 - k)[1 + 3\phi/2k]\}. \tag{31}$$

Eq. (31) implies that for $k < 1$ the decrease in the distribution coefficient leads to a decrease in the thereshold value of the growth velocity for the onset of interfacial deformations. Thus, the presence of the particles has a destabilizing effect.

The more general expression for the stability threshold, derived using the extremum principle in [7], can be written with c_∞ as the control parameter as

$$c_\infty = \frac{\mathcal{G}D/V + T_m\Gamma\alpha^2 D/V}{-m[(1 - k_{eff})/k_{eff}][(b + 1)/(b + 1 - k_{eff})]}, \tag{32}$$

where $\mathcal{G}$ stands for the conductivity weighted average thermal gradient, α is the wavenumber of the perturbation and

$$b = \frac{1}{2}[-1 - \sqrt{1 + 4\alpha^2}]. \tag{33}$$

The calculations of the critical values of the concentration c_∞^* and the wavenumber α^* for the case of a Pb-Sn alloy, with the numerical values of the physical constants obtained from Table 1 of reference [8], show that for $\phi \leq 0.01$ and $V = 1\mu m/s$,

$$c_\infty^* \approx 0.005695 - 0.01906\phi \qquad \text{and} \qquad \alpha^* \approx 14.69 - 11.81\phi. \tag{34}$$

Note that the presence of the particles leads to a decrease in the threshold value of the concentration and to an increase in the critical wavelength $\lambda^* = 2\pi/\alpha^*$.

CONCLUSIONS

We have investigated the modification of the Mullins-Sekerka instability by the presence of particles in a dilute binary mixture. The fluid is immersed with spherical particles which we assume to be positioned in a doubly-periodic array having an inter-particle distance that is at least four particle diameters to satisfy the requirement that the suspension be dilute [9]. It is shown that the particles cause the partition coefficient k to decrease (increase) if $k < 1$ ($k > 1$). A linear stability analysis of the planar interface reveals that the presence of the particles has a destabilizing influence, i.e the critical value for the onset of interfacial deformations is lowered due to the particles' presence in the melt. This destabilizing effect is attributed to the hindrance effect of the particles on the diffusion of solute away from the interface. This leads to the creation of a solute rich region adjacent to the solid front which is accompanied by an increase in the degree of supercooling. Hence the solute concentration level needed for instability is lowered. This decrease in the instability threshold is also associated with an increase in the critical wavelength.

We have also found that changes in the solutal diffusion coefficient and of the thermal gradient due the particles' presence have negligible effects on the results to the order at which the calculations are carried out (i.e. $\varnothing(\phi)$).

ACKNOWLEDGEMENTS

It is a pleasure to acknowledge helpful discussions with Professor D.M. Stefanescu and his students at the Solidification Laboratory, University of Alabama. This work has been supported by a grant from the National Science Foundation (DMS-9700380).

REFERENCES

[1] K.A. Jackson and B. Chalmers, J. Appl. Phys. **29**, 1178 (1958).

[2] A.E. Corte, J. of Geological Res. **67**, 1085 (1962).

[3] C. Korber, G. Rau, M.D. Cosman, and E.G. Cravalho, J. Crystal Growth **72**, 649 (1985).

[4] D.R. Uhlmann, B. Chalmers and K.A. Jackson, J. Appl. Phys. **67**, 1085 (1964).

[5] A.A. Chernov, D.E. Temkin and A.M. Mel'nikova, Sov. Phys. Crystallogr. **21**, 369 (1976); ibid., **22**, 656 (1976).

[6] D. Shangguan, S. Ahuja and D.M. Stefanescu, Met. Trans. **23A**, 669 (1992).

[7] R.F. Sekerka, J. Crystal Growth **3,4**, 71 (1968).

[8] S.R. Coriell, M.R. Cordes, W.J. Boettinger and R.F. Sekerka, J. Crystal Growth **49**, 13 (1981).

[9] M. Ungarish, Hydrodynamics of Suspensions (Springer-Verlag, Berlin, 1993), p.10.

Unconditionally gradient stable time marching the Cahn-Hilliard equation

David J. Eyre*

May 26, 1998

Abstract

Numerical methods for time stepping the Cahn-Hilliard equation are given and discussed. The methods are unconditionally gradient stable, and are uniquely solvable for all time steps. The schemes require the solution of ill-conditioned linear equations, and numerical methods to accurately solve these equations are also discussed.

1   Introduction.

Numerical time marching schemes for the Cahn-Hilliard equation will be discussed in this paper. The Cahn-Hilliard equation is a leading model in theoretical materials science, and the efficient numerical solution of the equation is needed to understand its dynamics. The equation is a fourth order nonlinear parabolic partial differential equation,

$$\frac{\partial u}{\partial t} = \Delta \left[-\varepsilon^2 \Delta u + f(u) \right] \tag{1}$$

where $u = u(\vec{x}, t)$ and

$$0 < \varepsilon \ll 1, \quad \vec{x} \in \Omega, \quad t > 0. \tag{2}$$

The nonlinear term taken here will be $f(u) = u^3 - u$, but the properties of the numerical methods hold equally for other bistable nonlinear terms. The operator Δ is the spatial laplacian. The linear part of $\Delta f(u)$ is $-\Delta u$, and this term is responsible for the interesting dynamics of the Cahn-Hilliard equation including the instability of constant solutions near $u = 0$. The nonlinear term of $\Delta f(u)$ is Δu^3, and this term stabilizes the flow. The Cahn-Hilliard equation was derived in [2] and numerical issues concerning the equation were surveyed in [3].

The natural and average conserving boundary conditions for the system are

$$\nabla u \cdot \vec{n} \,|_{\vec{x} \in \partial\Omega} = \nabla(\Delta u - f(u)) \cdot \vec{n} \,|_{\vec{x} \in \partial\Omega} = 0 \tag{3}$$

where $\vec{n}$ is the outward normal on the boundary of Ω. These boundary conditions leave the spatial average of $u(\vec{x}, t)$ invariant in time. The Cahn-Hilliard equation dissipates energy, i.e.

$$\frac{d\mathcal{E}(u)}{dt} \leq 0 \tag{4}$$

---

*Department of Mathematics, University of Utah, Salt Lake City, UT 84112, eyre@math.utah.edu

where

$$\mathcal{E}(u) = \int_\Omega \left[\frac{\varepsilon^2}{2}|\nabla u|^2 + F(u)\right] dx \tag{5}$$

and $f(u) = F'(u)$. Therefore, for all positive times, $F(u(t)) \le F(u(0))$.

The Cahn-Hilliard equation is notoriously difficult to solve numerically [3] because the equations are stiff due to both the biharmonic operator and the nonlinear operator. Additionally, across the spatial interfaces, the solution undergoes an $O(1)$ change over an $O(\varepsilon)$ interval. To accurately resolve these interfaces a fine discretization of space is required.

Therefore, most time marching schemes require one to use time steps that are many orders of magnitude smaller than the fastest times scales of the equation. For example, if Euler's method is employed to solve these equations with the given scaling and with $\varepsilon = 0.01$, then one is forced to take time steps on the order of 10^{-7} while the fastest interesting dynamic time scales are on the order of 10^{-4}. Therefore, thousands of time steps are required to see even the fastest dynamics. Other dynamic time scales are much longer, making explicit numerical schemes expensive.

Other explicit and most semi-implicit schemes have similar restrictions on their time steps. Alternatively, fully implicit schemes require that one solve a coupled system of nonlinear equations. If one chooses the time step for these schemes to be too large, the nonlinear equations have multiple solutions. Only one of these solutions is correct, and the spurious solutions confuse nonlinear equation solvers, like Newton's Method. Consequently, these solvers will not converge to any solution. Finally, it has been shown [7] that only a small set of fully implicit schemes are gradient stable in the sense that they have a discrete energy that decreases from one time level to the next. Furthermore, all of these schemes are conditionally gradient stable meaning that they only have a discrete gradient structure for small enough time steps. The allowed time steps are roughly one order of magnitude larger than the maximum stable time step allowed by Euler's method.

The goal of this paper is to apply a new semi-implicit method [4] to the Cahn-Hilliard equation that resolves the problems associated with both the stiffness and solvability. The methods are unconditionally gradient stable, which means that the equations have a discrete energy that decreases from one time level to the next for all positive time steps. The schemes also possess the same equilibria as the dynamic equations. The methods are also uniquely solvable for all positive time steps.

The methods require the solution of a ill-conditioned set of linear equations, and this paper also gives methods for accurately solving these systems of equations.

It should be emphasized that while the methods will allow one to take arbitrarily large time steps and not become unstable, the accuracy of the numerical solution will depend on choosing a small enough time step to resolve the dynamics.

2 Gradient stable time stepping schemes.

For simplicity, the time stepping method will be presented as a finite difference scheme in two spatial dimensions. One could use the time stepping with other spatial discretizations and in other spatial domains.

Using finite differences, the Cahn-Hilliard equation can be discretized with centered approximations to the spatial derivative operators on $\Omega = [0, 1] \times [0, 1]$. Let u_{ij}^n be the discrete approximation of $u(ih, jh, nk)$, where the time step is k, the spatial discretization is $h = 1/m$, and $m = O(\varepsilon^{-1})$. Then the standard centered difference approximation of the laplacian is

$$h^2 \Delta_h u_{ij}^n = u_{i+1,j}^n + u_{i-1,j}^n + u_{i,j+1}^n + u_{i,j-1}^n - 4u_{i,j}^n. \tag{6}$$

This operator will be used to approximate the continuous laplacian on the discrete grid.

The key to achieving the desired stability and solvability features of the time marching scheme is to divide the nonlinear terms into a stabilizing term and into a growth term, and then to separate those terms across the time step. The most logical way to accomplish this is with the time marching scheme

$$\frac{u_{ij}^{n+1} - u_{ij}^n}{k} = -\varepsilon^2 \Delta_h^2 u_{ij}^{n+1} + \Delta_h (u_{ij}^{n+1})^3 - \Delta_h u_{ij}^n. \tag{7}$$

To find u_{ij}^{n+1} via this scheme, one needs to solve the nonlinear equations

$$\left(1 + k\varepsilon^2 \Delta_h^2\right) u_{ij}^{n+1} - k\Delta_h (u_{ij}^{n+1})^3 = (1 - k\Delta_h) u_{ij}^n \tag{8}$$

at every grid point. This splitting satisfies the conditions presented in [4] for an unconditionally gradient stable time marching scheme.

If the grid points are gathered into a vector by a lexicographic ordering of the nodes and $\mathbf{U}^n \in \mathbf{R}^{m^2}$ is the solution at $t = nk$, then the difference equations read

$$\left(I + k\varepsilon^2 A_h^2 - kA_h D(\mathbf{U}^{n+1})^2\right) \mathbf{U}^{n+1} = (I + kA_h) \mathbf{U}^n \tag{9}$$

where $D(\mathbf{U})$ is a diagonal matrix with the vector $\mathbf{U}$ placed element-wise along the main diagonal. The matrix A_h is the ordered discrete laplacian, it has one zero eigenvalue and its remaining eigenvalues are negative, and it has the familiar block tridiagonal form.

The equations in (9) are a large coupled set of nonlinear equations. Numerical experiments in one spatial dimension show that Newton's method converges in two or three iterations for time steps on the order of the fastest time scales of the dynamics. The convergence requires four or five iterates for time steps less than one, and p additional iterates when $k = 10^p$ for $p > 0$.

While this nonlinear splitting is reasonable and has the smallest local truncation error of the one step methods that are gradient stable, the following splitting of the stabilizing and contracting terms leads to a set of *linear* equations that must be solved at every time step. This time marching scheme is

$$\frac{u_{ij}^{n+1} - u_{ij}^n}{k} = -\varepsilon^2 \Delta_h^2 u_{ij}^{n+1} + \Delta_h \left[(u_{ij}^n)^2 u_{ij}^{n+1}\right] - \Delta_h u_{ij}^n. \tag{10}$$

After ordering the nodes, the difference equations in (10) read

$$\left(I + k\varepsilon^2 A_h^2 - kA_h D(\mathbf{U}^n)^2\right) \mathbf{U}^{n+1} = (I - kA_h) \mathbf{U}^n. \tag{11}$$

Notice that the only difference between this method and the previous method is that the nonlinear matrix is evaluated at the previous time step. This scheme is also unconditionally gradient stable, and the local truncation error for this method is the same as for the previous method, $O(k)$. Furthermore the equations are uniquely solvable for all time steps. Therefore since the equations are linear, this method is the recommended scheme.

From this scheme, it is easy to see that the solution will remain bounded, and that the scheme is uniquely solvable for all time steps. The solvability property follows because the difference operators on the left hand side of the matrix equation are all nonnegative definite. Therefore, the matrix equation has strictly positive eigenvalues for all positive time steps. The boundedness follows from comparing the operator on the left hand side with the operator on the right hand side of the equation. Approximating the terms in $D(\mathbf{U})^2$ by their smallest value, one can see that the long wavelength perturbations cannot grow unboundedly because the eigenvalues of the operator on the right hand side are smaller than one in absolute value and the eigenvalues of the operator on the left hand side are larger than one. Similarly, short wavelength perturbations cannot grow unboundedly because their growth is strongly damped by the discrete biharmonic operator, A_h^2.

3 Solving the linear equations.

Ideally when working with discretizations of the laplacian, it makes sense to use FFT based methods to invert the operators because the FFT basis functions are eigenvectors of the difference operators. FFT methods require that the operator be linear and constant coefficient. Unfortunately, while the equations in (11) are linear, they are not constant coefficient equations and, therefore, FFT based methods cannot be used for the inversion.

Either direct or iterative methods could then be used to solve the linear equations. However, inverting this system is difficult because the equations are ill-conditioned. The ill-conditioning arises from the discretization of the biharmonic operator $-\Delta^2$ that is represented by the matrix A_h^2. The eigenvectors of this matrix are sinusoidal, and the large eigenvalues of this matrix scale like m^4 where m is the wave number of the eigenvector and the spatial discretization parameter. The small eigenvalues of A_h are zero, so the small eigenvalues of the update matrix are near one. Therefore, the condition number of the update matrix is $O(\varepsilon^2 m^4)$. For large enough m to make the calculations interesting, this condition number makes naive matrix solution methods unusable.

This section will present a method to precondition the linear equations in order to greatly reduce the condition number of the original matrix. Once the preconditioner is described, iterative methods are given to solve the transformed linear system of equations.

3.1 Preconditioning the system.

Nearly all matrix solution methods fail miserably when the system of equations they are trying to solve is poorly conditioned. Roundoff error accumulates to the point that it dominates the computed answer, and the effect is that the "solution" computed by the chosen method is useless. One logical choice to resolve this problem is to transform the original system of linear equations into a new system of linear equations that have better conditioning, and then to solve the problem in the new transformed variables. This approach, called preconditioning, will be applied to solve the update equations defined in (11). The approach given here results in a matrix whose condition number is nearly one, and is therefore, nearly optimal.

Notice that the only term with nonconstant coefficients in the linear equations defined in (11) are the elements of the diagonal matrix $D(\mathbf{U}^n)^2$. This matrix has nonnegative entries, and once the solution of the Cahn-Hilliard equation has a layered structure, the matrix is nearly the identity everywhere in space except within the layers themselves. Therefore the equations in (11) are *nearly* constant coefficient equations. This suggests that a constant coefficient approximation to the update matrix would make a good preconditioner.

Let $\hat{D} = d^2 I$ be the constant approximation of $D(\mathbf{U}^n)^2$, and define the preconditioner to be

$$M = I + k\varepsilon^2 A_h^2 - kd^2 A_h. \tag{12}$$

The results of numerical tests indicate that taking d^2 equal to the mean of the elements of $D(\mathbf{U}^n)^2$ and taking $d^2 = 1$ are both reasonable choices. It will now be shown that this preconditioner is nearly optimal.

Using the preconditioner, the original equations are replaced with a transformed system of linear equations of the form

$$M^{-1}\mathcal{A}\hat{x} = \hat{b} \tag{13}$$

where

$$\mathcal{A} = I + k\varepsilon^2 A_h^2 - kA_h D(\mathbf{U}^n)^2, \tag{14}$$

and

$$\hat{b} = M^{-1}\left(I - kA_h\right)\mathbf{U}^n, \text{ and } \hat{x} = M^{-1}\mathbf{U}^{n+1}. \tag{15}$$

These condition number of equations is now analyzed.

The effects of this preconditioner on the condition number of the problem are easily seen. First, since A_h has one zero eigenvalue corresponding to a constant eigenvector, the preconditioned matrix will have a smallest eigenvalue nearly equal to 1. To find the largest eigenvalue of the preconditioned matrix, the equations $M^{-1}\mathcal{A}$ are calculated as

$$\left(I + k\varepsilon^2 A_h^2 - kd^2 A_h\right)^{-1}\left(I + k\varepsilon^2 A_h^2 - kA_h D(\mathbf{U}^n)^2\right)$$

$$= \left(I + k\varepsilon^2 A_h^2 - kd^2 A_h\right)^{-1}\left(I + k\varepsilon^2 A_h^2 - kA_h D(\mathbf{U}^n)^2 - kd^2 A_h + kd^2 A_h\right)$$

$$= I + \left(I + k\varepsilon^2 A_h^2 - kd^2 A_h\right)^{-1} kA_h \left(d^2 I - D(\mathbf{U}^n)^2\right).$$

To estimate the two norm of this latter matrix note that

$$\|I + \left(I + k\varepsilon^2 A_h^2 - kd^2 A_h\right)^{-1} kA_h \left(d^2 I - D(\mathbf{U}^n)^2\right)\|$$

$$\leq 1 + \|\left(I + k\varepsilon^2 A_h^2 - kd^2 A_h\right)^{-1} kA_h \left(d^2 I - D(\mathbf{U}^n)^2\right)\|$$

$$\leq 1 + \|\left(I + k\varepsilon^2 A_h^2 - kd^2 A_h\right)^{-1} kA_h\| \, \|d^2 I - D(\mathbf{U}^n)^2\|$$

To analyze the norm of the first system, notice that the eigenvectors of the matrix

$$\left(I + k\varepsilon^2 A_h^2 - kd^2 A_h\right)^{-1} kA_h$$

correspond to the eigenvectors of A_h. Therefore, if X is an eigenvector of A_h with eigenvalue $\lambda < 0$, then

$$\left[\left(I + k\varepsilon^2 A_h^2 - kd^2 A_h\right)^{-1} kA_h\right] X = \left(\frac{-k\lambda}{1 + k\varepsilon^2\lambda^2 - kd^2\lambda}\right) X. \tag{16}$$

The eigenvalues are bounded above by

$$\frac{-k\lambda}{1 + k\varepsilon^2 \lambda^2 - kd^2 \lambda} < \frac{1}{d^2},$$

and this value bounds the two norm of the first matrix as well since symmetric matrices two norm equals their largest eigenvalue. Suppose that d^2 is chosen to satisfy $(u_{ij}^n)^2 \leq 2d^2$, then combining this with the last inequality gives

$$\| \left(I + k\varepsilon^2 A_h^2 - kd^2 A_h \right)^{-1} kA_h \| \, \|d^2 I - D(\mathbf{U}^n)^2\| < \frac{1}{d^2} \max_{ij} |d^2 - (u_{ij}^n)^2| < 1. \tag{17}$$

Furthermore, the analysis suggests that an optimal choice for d^2 is to satisfy the minimax criteria

$$\min_d \max_{ij} \frac{1}{d^2} |d^2 - (u_{ij}^n)^2| \tag{18}$$

at every time step. Thus, the condition number of the product $M^{-1}\mathcal{A}$ is less than two which is a nearly optimal preconditioning.

3.2 Fast inversion of the preconditioner.

For the preconditioner to be useful, it must be capable of being inverted quickly. The equations of (12) are constant coefficient linear equations, and can be quickly solved using FFT methods. These methods are described in most textbooks on the numerical solution of partial differential equations (for example, see [1]). Briefly, FFT based methods exploit the fact that the basis functions of the FFT are eigenfunctions of the discrete laplacians. More precisely, the cosine basis functions that are used in a discrete cosine transformation are eigenfunctions of the Cahn-Hilliard equations with the boundary conditions given in (3). The eigenvalues corresponding to these eigenfunctions for the finite difference method presented here are

$$\lambda_{kl} = -4m^2 \left(\sin^2 \frac{k\pi}{2m} + \sin^2 \frac{l\pi}{2m} \right) \tag{19}$$

The eigenvectors of the preconditioner matrix (12) are also discrete cosine functions, and the eigenvalues corresponding to these eigenvectors are the sum of the eigenvalues of the individual matrices, or

$$(I + k\varepsilon^2 A_h^2 - kd^2 A_h)x_{kl} = (1 + k\varepsilon^2 \lambda_{kl}^2 - kd^2 \lambda_{kl})x_{kl} \tag{20}$$

Therefore, to solve a system of the form $Mx = b$ where M is the preconditioner requires three steps;

- Compute the fast cosine transformation $\hat{b}$ of b.

- Divide the elements of $\hat{b}$ by the eigenvalues of M.

- Compute the inverse fast cosine transformation of the result.

This fast solution technique will be used as the basis for the iterative solution of the original system.

3.3 Iterative solutions of the preconditioned system.

Two logical choices for an iterative method to solve the preconditioned are now presented. The first method is an elementary iterative method that is presented to demonstrate the iterative process. For this method, recall the preconditioner matrix M, define N as follows

$$N = kA_h \left(d^2 I - D(\mathbf{U}^n)^2 \right),$$

and define a sequence of approximations $\mathbf{U}_\ell^{n+1}$ to $\mathbf{U}^{n+1}$ as

$$M\mathbf{U}_{\ell+1}^{n+1} = (I - kA_h)\,\mathbf{U}^n - N\mathbf{U}_\ell^{n+1}$$

where ℓ is the iteration parameter. From the definitions, notice that

$$M + N = I + k\varepsilon^2 A_h^2 - kA_h D(\mathbf{U}^n)^2,$$

so if $\mathbf{U}_{\ell+1}^{n+1} = \mathbf{U}_\ell^{n+1}$ then $\mathbf{U}_{\ell+1}^{n+1} = \mathbf{U}^{n+1}$ because the linear equations have a unique solution.

The iterative scheme is guaranteed to converge if the spectral radius of the iteration matrix $M^{-1}N$ is less that one [5]. The spectral radius of $M^{-1}N$ was bounded below one in equation (17) provided a reasonable choice of d^2 is made. Thus the iteration procedure will converge.

The first method demonstrates a simple iterative solver that is guaranteed to converge with reasonable choices of the parameters that are available. However, far more sophisticated iterative schemes exist that will converge faster than this scheme. The most logical choice to solve the preconditioned system presented in this paper is the method known as conjugate gradients squared (CGS). A formal presentation of this method is beyond the scope of this paper. Its details are presented in [6], and well documented and tested code for CGS is available from *www.netlib.org*.

The convergence rate for CGS applied to this system when d^2 is chosen via the minimax criteria (18) is available from analytic results. These results indicate that the solution will gain one decimal digit of accuracy for every 2.1 iterations. Therefore, 16 to 20 iterations is enough to compute the solution to single precision accuracy, and computing many more iterates is wasteful because the accuracy of the scheme, $O(k)$, doesn't significantly more work on the linear system. Since each CGS iteration requires only $O(m^2)$ effort, the overall effort required to solve the linear system is dominated by the effort to perform the FFTs and is $m^2 \log m$. The system itself has m^2 unknowns, therefore, the method is nearly optimal. Furthermore, since the time marching method is unconditionally stable, one can take large steps and effectively solve the Cahn-Hilliard equation over long time intervals.

References

[1] W. F. AMES, *Numerical Methods for Partial Differential Equations*, Academic Press, New York (1977), pp. 304–307.

[2] J. W. CAHN, Acta Met., 9 (1961), pp. 795–801.

[3] C. M. ELLIOTT, in Mathematical Models for Phase Change Problems, J. F. Rodrigues, ed., Birkhäuser Verlag, Basel, 1989, pp. 35–73.

[4]   D. J. EYRE, preprint.

[5]   G. GOLUB AND C. F. VAN LOAN, Matrix Computations, Johns Hopkins Press, Baltimore (1983), pp. 353–361.

[6]   P. SONNEVELD, SIAM J. Sci. Stat. Comput., 10 (1989), pp. 36-52.

[7]   A. M. STUART AND A. R. HUMPHRIES, SIAM Rev., 36 (1994), pp. 226-257.

DIRECT METHOD OF MICROSTRUCTURE MODELING

A.A.AIVAZOV*, N.V.BODYAGIN*, S.P.VIKHROV**
*MT Faculty, The Moscow Institute of Electronic Technology, Moscow, Russia,
aivazov@mictech.zgrad.su
**CR Faculty, The Radiotechnical Academy, Riazan, Russia

ABSTRACT

There discussed some possibilities basic concept of the nonlinear system theory and methods of modeling the microstructure of solids. It is demonstrated, that the substance distribution on the surface and in the volume of some materials possesses a determined chaotic character.

INTRODUCTION

The most important solid-state theory problem lies in the fact that so far the laws of solids state structure formation and subsequent evolution are unclear and, as a result, adequate methods of description of a correlation between the structure and physical-chemical properties of materials and prerequisites to their growth are now lacking.

This determines the necessity to develop new approaches to structure modeling and structure formation processes. One possibility is that methods of the nonlinear dynamics are used. This assumption seems justified since the substance in the process of solidification displays the properties of a nonlinear, self-organized system [1].

THEORY

The main concept of the experimental study of the self-organized system evolution implies that, the results of the system observations in the appropriate phase space present a geometric object, the dimensions of which are considerably smaller than those of the original space. Therefore, unravelling the dynamics of the systems involves observation of the evolution of the system attractor rather than of the entire infinite-dimensional original phase space [2].

The basic technique for the nonlinear system analysis is the embedding Takens method. With this metod, using it, it is possible to distinguish systems that show a chaotic movement in a space of low dimensional from systems with noise, and to measure invariant characteristics of the chaotic systems.

The technique is based on the idea that any system signal contains information on all the processes inside the system, since all the parts of a dynamic system are interconnected and can be viewed as a single whole. Therefore, system behaviour is decipherable through measuring any of its characteristics carried out at regular intervals. The data sequence obtained is treated on special algorithms, and there defined type and the dimensions of the attractor, the number of degrees of freedom, Lyapunov exponents and other dynamic parameters [2].

As on solidification the substance changes its properties both with space and time, two possibilities of research into growth dynamics are available:
- with respect to some characteristics of the substance *in situ* in terms of the theory of the Takens embedding method;
- with respect to the material structure that stores information on its previous temporal evolution. The first step towards restructuring the spatio-temporal phase space is the treatment of the purely spatial space series. This approach is quite legitimate if we study the "frozen" spatial disorder resulting, for example, from the surface of the materials. For this pure space series all the machinery developed for time series can be directly reformulated by substituting spatial coordinate x for time coordinate t. In this way one can calculate the fractal dimensions of the "space" attractor, evaluate the Lyapunov spatial exponents, construct a model, etc.

EXPERIMENT

For some materials technologies the instantaneous shot of the evolution process well-approximation, is the surface.

On this assumption, there have been measured the fractal dimensions (D) of the a-Si:H, GaAs, carbon, tungsten surfaces and other material surfaces on the surface profile, obtained with the help of the scanning tunneling microscopy and atomic power microscopy. The profile height was counted out from some level accepted as zero and was measured at discrete distances. The choice of the reconstruction parameters the time of delay, the number of measurement points was substantiated. Then the obtained data were processed with the help of the Grassberger-Procaccia algorithm, and the correlation dimension D and its dependence on the embedding dimension n were defined. The typical dependence of D on the resolution scale (r) is shown in fig.1 "a".

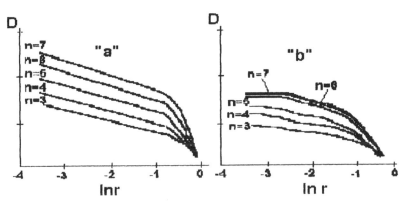

Fig.1. The dependence of D on the resolution scale (r) and the embedding dimension (n).

It is obvious that though D is not identified there is a linear portion. And the slope size in this portion is saturated as n increases. According to the approach [3], this means that the surface structure has a determined chaotic distribution in which degrees of freedom are algebraically independent, and the correlation length is much less that the size of the system. In addition to the embedding method, the data for the surface profile were examined by application of the function of distribution, the Fourier analysis, the function of average mutual information.

The results of the investigations for the materials listed are the following:

The surface of the a-Si:H grown in a glow discharge.

The dependence of D=f(n, ln r) on the substrate temperature in the material growth has a complex nature. In all cases there observed three different portions in the ranges Δln r: from -0,6 to 0, from -0,6 to -1,4, from -1,4 to -1,8. In all the portions there are areas of linear slope which is saturated with the embedding dimension of n=8. With increasing substrate temperature these portions become more marked.

The presence of the portions that have different scaling, means that the fractal set formed by the data for the profile of the surface in question, is non-uniform but three-scaled in this case - and surface formation is determined by three different mechanisms.

The surface of the carbon grown by pyrolysis.

On the diagram D=f(n, ln r) two portions having different maximums, are observed. Both the portions are characterized by the qualitatively identical behavior of D. In both the cases there are areas Δln r from -2 to 1.7 and from -3.5 to -3, respectively, in which the dependence D = f(n, ln r) is linear. With n=6 the slope D = f(ln r) in these portions is saturated.

The surface of the crystal chip of GaAs.

On the diagram D=f(n, ln r) there observed two portions having different maximums. In both the cases there are areas Δln r from -0.7 to -0.4 and from -3.2 to -2.5, respectively, in which the dependence D = f(n, ln r) is linear. For the first area the saturation of the slope D = f(ln r) occurs at n=6 and for the second one at n=7. Thus, as in the case of carbon, there observed a two-scaled fractal set, each of which has a determined chaotic character and is formed by two different mechanisms.

There has also been studied the dependence of D = f(n, ln r) on the scale of the portion, in which the measurements were carried out.

On the passivation of the GaAs surface by a diluted solution of NaOH it was found out that distribution ceases to be determined and possesses a casual character.

Tungsten.

The samples were prepared according by the following technique: the cutting of a monocrystal, the grinding and the polishing of the surface by application of electropolishing in the NaOH solution at the final stage. The character of the dependence of D=f(n, ln r) testifies that the surface structure has a casual character. After cleaning the surface by repeated scanning made by the electronic microscope, there observeda linear portion in the range Δ(ln r) from -2,3 to -1,9 on the dependence of D=f(n, ln r). The slope is saturated at n=6. It shows the determined character of the distribution and it is possible to speak about the "displaying" of the ordered surface structure.

We have compared the results obtained with already known experimental data for the values of fractal dimensions of the surfaces of various materials and for bulk samples of porous and amorphous materials. It is established that the value D is enclosed in the interval from 2 to 3 and has a fractional part. It is supposed that the fractal dimensions measured by physical properties, represents the average meaning of the local structure topology. The dependence of D= f(n, ln r), shown in fig.1 "b", is typical of it. Thus, the value D is the indicator of the chaotic distribution, which has a determined nature.

We have considered the possibility of determining the formation dynamics type by the system structure. By analogy with the Lyapunov exponents, describing the time system evolution λ_t, for the spatial structure the concept of the spatial Lyapunov exponents λ_r is being introduced. With the approach worked out in [4], it is established that the following interrelation exists between them:

$$\lambda_t / \lambda_r = C\, v, \tag{1}$$

where C is a positive constant, v is a value determined by the growth rate of the material and the atom diffusion on the growth surface. On the basis of this result the conclusion was made that the dynamics of material growth is defined by the determined spatial and temporal chaos. The analysis of already known experimental data confirms this conclusion.

On this basis the interrelation between the characteristics of the structure and the invariant parameters of its formation dynamics was established.

Relying on the result obtained it is possible to establish a qualitative interrelation between the structure parameters and the characteristics of the spatial-time dynamics.

The Lyapunov exponents for the spatial-time system $\lambda_{t,r}$ can be determined by the formula

$$\sum_i \lambda_{t,r} = \sum_i \lambda_{ri} + \sum_j \lambda_{tj}, \tag{2}$$

where $\sum \lambda_{ri}$ is the sum of the Lyapunov spatial exponents, $\sum \lambda_{tj}$ is the sum of the Lyapunov time exponents, i and j are the number of spatial and time parameters, respectively, i+j= $R_{t,r}$ - is the dimensions of the system phase space.

For the fractal dimension of the spatial-time chaos regime $D_{t,r}$ the following equation is true:

$$D_{t,r} \ge D_r, \tag{3}$$

where D_r is the fractal dimension of the spatial distribution.

The characteristics of the material structure and growth, based on the invariants of the nonlinear dynamics, can be of great practical value for modeling and control of growth processes.

By the dimension of the phase space and the Lyapunov exponents it is possible to judge the extent to which the substance is distant from the balance. It might be important in defining ways of material characteristics stabilization.

The local and global Lyapunov exponents, the topological entropy make it possible to determine the limits of the predictable mode of the system behaviour on different spatial and temporal scales.

MODEL OF AMORPHOUS STRUCTURE

On the basis of already known experimental data it is established that the fractal dimension of the volume structure a-Si:H has the value of 2<D<3. In accordance with above stated the results, it testifies that the a-Si:H structure represents the "frozen" determined chaos, i.e. it is not casual. A similar situation takes place in the case of some other noncrystalline materials.

According to this, there was constructed and analyzed a model of amorphous structure formation as the one-dimensional map

$$X_{n+1} = A + \lambda(A - |X_n|) \tag{4}$$

X_n and X_{n+1} are the energies of the connection between the n and n+1 atoms in the chain, respectively. X_{n+1} is the function of the deviation from its equilibrium value (A). The assumption that under any growth condition the interatomic connection tends to the steadiest configuration, corresponding to the crystal, is the physical basis of the model. From the dependence of atom connection energy of on temperature it was obtained that the value λ is inversely proportional to the temperature at which the formation of the chain occurs.

At $\lambda<1$ the sequence $\{X_n\}$, determined by (4), at any initial value of X, converges quickly to A, forming an ordered chain. At $\lambda=1$ in the sequence $\{X_n\}$ there appear bifurcations of doubling period. With λ further increase the number of bifurcations grows and the system passes into a chaotic mode, which corresponds to an amorphous state. Then the picture becomes more complicated: the chaos alternates with the intervals, where there is an ordering (fig.2, $\lambda=1.57$). This corresponds to theformation of ordered areas in the amorphous matrix in the form of average order, columns, pores, etc. The values of λ, at which the origin of the amorphous grid and ordering are observed, are in good agreement with the experimental data.

There was found a conformity between changes in the histogram of sequences $\{X_n\}$ at increasing values of λ and the evolution of the TO-mode for the a-Si:H sample, grown at consistently decreasing substrate temperatures.

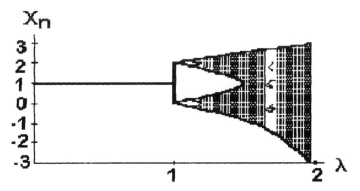

Fig.2. 1 Behaviour of the sequence $\{X_n\}$ at different values of λ.

SIMULATION PRINCIPLES

For the simulation of the microstructure and solid growth processes the following algorithm may be efficient:
- the first stage - experimental investigation of the surface or volume material structure by the embedding method;
- the second stage - numerical definition of the attractor type and fractal dimensions, the number of degrees of freedom, the Lyapunov exponents and other "dynamical parameters" of the structure;
- the third stage - nonlinear model building on the basis of information about the spatial distribution and characteristics of the spatial compact attractor in the phase space.

CONCLUSION

The stricture of the surface (a-Si:H, GaAs, C), representing the instantaneous "frozen" snapshot of growth processes, has determined the chaotic character. Proceeding from this, there formed a simulation principle microstructure and solid growth processes. The progress made in understanding [2] and controlling of the chaotic dynamic processes [5] lets us hope to solve the problem of obtaining reproducible stable and controlled solid characteristics.

REFERENCES

1. A.A.Aivazov, N.V.Bodyagin, S.P.Vikhrov, S.V.Petrov, J. Non-Cryst. Solids, **114,** 157 (1989).

2. H.D.I.Abarbanel, R.Brown, J.J.Sidorowich, L.S.Tsimring, Rev. of Mod. Phys. **4,** 1331 (1993).

3. L.S.Tsimring, Phys. Rev. B, **5,** 3421 (1993).

4. P.Grassberger, Phys. Scripta, **40,** 346, (1989).

5. T.Shinbrot, Adv. in Phys. **44,** 73 (1995).

Part II

Simulation of Microstructure Evolution

SIMULATION OF THE GROWTH OF HETEROSTRUCTURES

J.H. HARDING, A.H. HARKER
Dept. Physics & Astronomy, University College London, UNITED KINGDOM

ABSTRACT

The production and morphology of hetero-structures presents problems at a variety of length-scales. A common problem is the production and accommodation of stresses in the film due to mis-match. We shall first discuss examples of atomistic nucleation and growth at interfaces and the use of atomistic simulations to obtain parameters for rate-theory models of cluster and film growth. We shall then consider the effect of stress on growing films. In strained-layer semi-conductor systems, for example, the growth of small islands gives rise to stress distributions which differ strongly from those in continuous layers. Interesting strain effects are also observed in ceramics. We will discuss the relationship between stress and the growth and morphology of films, where effective medium models may be used to derive effective bulk properties for films with imperfections such as porosity and cracks.

INTRODUCTION

Heterostructures are both an area of great technological interest and an excellent example of the problems that arise when one is compelled to model a system at a variety of length-scales. The range of technological areas where heterostructures occur defies summary; semiconductor/semiconductor structures in photonic devices [1] and quantum dots [2]; metal/ceramic in coatings (and of course inadvertently in corrosion problems); ceramic/ceramic in ferroelectrics [3].

The growth of heterostructures can be considered at various levels of detail. First, there is the level of atomic mechanism. Here there are two approaches; direct simulation by molecular dynamics or Monte Carlo and the use of rate theories coupled with the calculation of individual processes. The strength of the direct simulation approach is that all processes are automatically included; its weakness is that limitations of computer time make the coatings unreasonably thin or the deposition rates used in the simulation impossibly high [4]. The methods based on rate equations avoid this problem but have the alternative difficulty that it is now necessary to identify all the relevant growth processes and incorporate them in the model.

Second, there is the problem of strain. A hetero-interface involves two incommensurate lattices and thus misfit is unavoidable. In real systems this sets up strains in both lattices. For thin films on top of a substrate, the thin film accommodates the strain up to a critical thickness; thereafter the misfit is accommodated by dislocations and the strain is relaxed. The calculation of this thickness is discussed in [5, 6]

THE ROLE OF SIMULATION

There are three main modes of growth possible when a material is deposited on a surface: layer by layer (Frank–van der Merwe) growth, island (Volmer–Weber growth) and layer then island (Stranski–Krastanov growth). These modes have been discussed in detail by Venables et al [8, 9]. Although growth modes are characterised by the thermodynamics

Table I: Values for the rate theory parameters E_a and E_d. Experimental estimates in round brackets [8, 11, 12] and previous calculations [13, 14] in square brackets.

Parameter	Ag/NaCl	Au/NaCl	Ag/NaF	Au/NaF	Ag/CaF$_2$
E_a(eV)	0.27, [0.27], (0.41)	0.15, [0.69], (0.49)	0.26	0.18, [0.59], (0.63)	0.36
E_d(eV)	0.15, [0.09], (0.19)	0.07, [0.22], (0.16)	0.24	0.14, [0.08], (0.08)	0.34 (0.4–0.5)
$E_a - E_d$ (eV)	0.12, [0.18] (0.22)	0.08, [0.47] (0.33 ± 0.02)	0.02	0.04, [0.51]	0.02

of the system, growth away from equilibrium requires a detailed study of the kinetics. One of the simplest and most successful is the pair-binding model of Venables [9]. Here, the maximum value of the nucleation density, N_x, for 2-D growth in the complete condensation limit is given by an equation of the form

$$N_x \propto R^p \exp\{(E_i + iE_d)/(i+2)kT\} \tag{1}$$

where R is the deposition rate (often denoted the flux F in the recent literature), T is the deposition temperature and E_d the surface diffusion activation energy for the adatom. The quantity $p = i/(i+2)$, where i is the critical nucleus size, is calculated self-consistently within the model. The model uses the lateral binding energy of arbitrary 2-D clusters, $E_j = b_j E_b$ (where b_j is the number of bonds and E_b is the binding energy of a pair of adatoms on adjacent sites). It evaluates i as that cluster size j (and configuration) which results in the lowest nucleation rate and density at the deposition temperature consistent with the constraints of the model. The model also allows for incomplete condensation, using a more complete expression than equation (1) and can deal with 3-D islands, as well as 2-D monolayer clusters [9]. The description of incomplete condensation requires knowledge of the adatom adsorption energy, E_a since the expression for the cluster density (in the extreme incomplete limit) is then

$$N_x \propto R^p \exp\{(E_i + (i+1)E_a - E_d)/kT\} \tag{2}$$

and $p = i$. The value of N_x, on a perfect, clean, substrate is a sensitive test of E_d and E_b, and, at higher temperatures, also of E_a. Higher values of E_b prolong the lower critical sizes, and higher nucleation density, to higher temperatures.

The calculation of the disposable parameters of such a model is a problem that can be tackled by computer simulation. Results for growth of noble metals on alkali halides have recently been published [10] and are shown in Table I.

Much has been achieved in understanding the nucleation and growth of metal clusters on substrates using simple models. However, the nucleation and growth of metal clusters on ionic substrates is a complex process involving both the trapping effects of line steps and surface point defects, and the participation of other mobile species, dimers, trimers and other even larger clusters. This illustrates both the strengths and the weaknesses of the rate theory approach. Provided that the implicit mean-field theory approximation is accepted (for a discussion of how adequate this approximation is see [15]) it is possible to construct a direct connection between individual processes and the nucleation and growth of coatings. The weakness is that one must know what all the processes are.

In the simulation of nucleation, or of very thin films, one can ignore the geometrical problem of fitting two different lattices together. The layer being deposited will always adjust to the requirements of the substrate. Once the deposited layer is greater than a few atomic layers, this is no longer so. For hetero-systems a few layers thick it is still necessary to balance optimising the local coordination at the interface and the energy built into the film by the stresses. This can be investigated by direct computer simulation (see, for example, [16, 17]). These authors considered various coincidence-site lattices (for an introduction to coincidence-site lattice theory see [18]) and showed that the lattice with lowest misfit was not always the only one observed. Such considerations are usually ignored in electronic structure calculations which impose strict epitaxial matching with small interface lattices because of computational limitations. This is seen, for example, in all calculations performed on the MgO/Ag interface (see [19] for a review).

THE CONTINUUM LIMIT

We next consider the calculation of strains in heterointerfaces. A fundamental issue (recently reviewed in [21, 22]) is the critical distance when the energy locked up in the stresses in the films is sufficient to drive the nucleation of interface dislocations. These relieve misfit stresses (and also incidentally permit a better matching of the local structures at the interfacial boundary). These stresses have often been calculated using simple analytical approximations (see, for example [20]). However, recent work using a two-dimensional finite element method [21] has shown the inadequacy of many of these approaches. This work shows the complex stress states that can arise for epitaxial layers depending on the thickness of the layer and the width of the layer relative to the width of the substrate.

As an example of the complexities that can arise and of their relevance to problems of great current technical interest we consider the growth of gallium nitride/aluminium nitride heterostructures. A typical solid state laser structure is shown in Figure 1: it is based on a sapphire substrate, and is a multi-layer structure at the heart of which is a sequence of alternating layers of $In_{0.15}Ga_{0.85}N$ and $In_{0.02}Ga_{0.98}N$. A typical structure is about 4 μm wide, 550 μm long, and 2 μm high, but the quantum well structure occupies only about 30 nm of that height. A simple estimate of the stresses in the structure may be obtained from the lattice mismatch strains, on the assumption that these are constant across the width of the structure. Although one might guess that the height/width ratio would be enough to ensure uniformity across the width of this quantum well structure, finite element calculations (Figure 2) show that there is a region about 150 nm wide at each edge of the structure in which the stress differs significantly from that at the centre of the structure. Given the stress dependence of the electronic structure of GaN ([23]) the photoluminescence energy of the material in the quantum well region may be expected to vary by about 0.1 eV over this region.

These stress calculations were performed under an assumption of isotropic elastic behaviour. In reality, the laser structures may be made of material with hexagonal symmetry, strained on the (0001) interface, and as well as stresses there are piezoelectric effects which should be taken into consideration.

BUILDING FILMS AT LONGER LENGTH-SCALES

Up to now we have considered only two length-scales; the atomic scale and the continuum limit. A brief mention of dislocations in previous sections is a reminder that mi-

III-Nitride Laser Structure

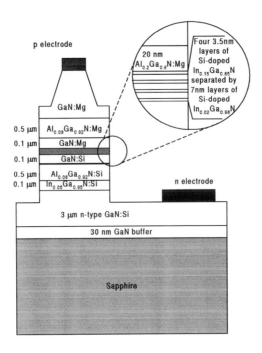

Figure 1: Schematic diagram of a quantum well structure laser.

crostructures, with length scales on the order of microns, are frequently important. Here a common approach is the use of phenomenological models. In building up films, the commonest models used are the Solid-on-Solid and Terrace-Site-Kink models (see [24] for a brief introduction), which are often built on a mesh and almost always approach the problem by identifying a process or processes that are believed to underlie the observed behaviour. These are formulated as rules for computation and the mesoscale process is followed by computer. This strategy, expressed in such general terms, is of wide application. The models, inevitably, tend to be strongly situation-dependent. In all cases there are problems of the meaning of the parameters in terms of fundamental mechanisms, which are often quasi-atomistic processes. The 'blocks' of the model are deemed to be primitive models of atoms or groups of atoms.

This need not be so; the same basic idea is usable on an intrinsic micron scale, as can be seen in the modelling of plasma-sprayed coatings [25]. Essentially, the plasma-spray process is the projection of small particles of semi-molten or molten material of micron size onto a substrate using a plasma gun. Here basic unit is the individual particle which hits the coating and splashes. The coating is made up of the resulting splats. The growth rules are

**Stress (Sxx) for 4 micron width
(n.b. only part width shown)**

GPa

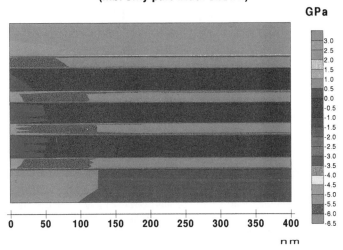

	3.0
	2.5
	2.0
	1.5
	1.0
	0.5
	0.0
	-0.5
	-1.0
	-1.5
	-2.0
	-2.5
	-3.0
	-3.5
	-4.0
	-4.5
	-5.0
	-5.5
	-6.0
	-6.5

0   50   100   150   200   250   300   350   400

nm

Figure 2: Stress in the quantum well region of a solid-state laser. The stress plotted is σ_{xx}, the x direction being the horizontal in the diagram, and only the stress in part of the 4 μm width of the structure is shown.

more complex than before since the shape of the basic units changes drastically during the process. For a wide range of spraying conditions, the resulting coatings are full of pores and cracks. This is not a defect; for thermal barrier coatings such a microstructure helps achieve low thermal conductivity and good resistance to residual stress. The dependence of the microstructure on the process conditions can be modelled and examples are shown in Figure 3.

DEALING WITH THE MICROSTRUCTURE

A vital connection to make is that between the effective bulk properties required in the calculations discussed above and the structures discussed in the previous section. In the end, unless one proposes to perform a simulation at all length-scales, retaining all the information gained at each length-scale, some kind of averaging must be performed. A long-established method for doing this is effective medium theory. The essential idea of such a theory is to calculate the average response $< r >$ to a stimulus $< f >$ using the

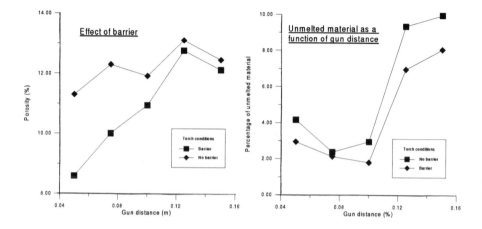

Figure 3: Effect of adding an air jet to prevent overheating of the substrate during plasma spraying(left); Effect of increasing the distance of the substrate from the plasma-spray gun; large amounts of unmelted material are undesirable as the resulting coatings do not adhere well (right).

idea of a generalised susceptibility S. The basic equation is thus $< r >= S < f >$ (or equivalently $< f >= C < r >$ using the generalised stiffness C). In general, $< r >$, $< f >$ are vectors and so S, C are tensors. A general discussion is given by Batchelor [26]; specific examples go back to Maxwell [27], Lorenz [28] and Rayleigh [29]. Here we consider only self-consistent theories.

Self-consistent theories of effective properties of materials attempt to account for the fact that, for a given inclusion, the surrounding material is not the original matrix but is already modified by the presence of other inclusions. Care is often required in the use of such theories, since they may not generalise correctly to the case of multiple media. A simple theory that can be so generalised is the symmetric effective medium theory [30]. Such a theory also displays a percolation threshold [31] but this may not appear in the right place (see the discussion for the elastic case [32, 33]).

A popular approach to the effective medium properties of materials is Eshelby theory [34]; for a recent review see [35]. The effective conductivities and elastic properties of materials containing ellipsoidal inclusions may all be expressed in terms of tensors, dependent only on the axes of the ellipsoids in the case of conductivity but also depending on Poisson's ratio in the case of elastic constants. The tensors required have been tabulated by Mura [36].

60

In what follows we shall consider the example of a porous material. In this case, the effective conductivity is given by

$$\kappa^{\text{eff}} = \kappa_0 \left\{ 1 - \phi[[(1 - \phi)M + \phi] - 1]^{-1} \right\}^{-1} \tag{3}$$

where κ_0 is the conductivity of the pure material, ϕ the fraction of voids and M the Eshelby tensor. The same form of expression holds for elastic properties, except that now a fourth-order Eshelby tensor is required. An equivalent expression has been derived by Ferrari [37].

A somewhat different approach is required for cracked materials since the volume fraction of cracks is zero. We adapt a differential scheme as follows. The problem of determining effective elastic moduli $< C >$ and compliances $< S >$ from the equivalent quantities for the component phases may be formally solved [38] by writing

$$< C > \; = C^{(1)} + \phi(C^{(2)} - C^{(1)})A \tag{4}$$
$$< S > \; = S^{(1)} + \phi(S^{(2)} - S^{(1)})B \tag{5}$$

where the fourth-rank tensors A, B relate the average strain over the inclusions to the overall average (or make the similar conversion for stress), and depend on the shapes of the inclusions. Applications of this treatment to ellipsoidal cracks are discussed by Hashin [39].

The case of a material containing both pores and cracks is more complex. We have considered two approaches. The first is that proposed by Ferrari [37] using the Wu [33] characteristic tensor

$$T^{(i)} = (I + M^{(i)}C^{(i)^{-1}}(C^{(i)} - C^{(0)}))^{-1} \tag{6}$$

Here the superscript (0) refers to the matrix material; the model treats matrix and inclusions differently. The effective elastic constants are

$$C^{\text{eff}} = C^{(0)} + \Sigma_i'\phi_i(C^{(i)} - C^{(0)})T^{(i)} \left[\phi_0 + \Sigma_j'\phi_j T^{(j)} \right]^{-1} \tag{7}$$

where the primes denote the omission of matrix terms. The other scheme uses a mixture method, based on the procedure of Pedersen and Withers [40]. In the case of elastic materials, we first define a reference material with elastic properties $C^{(0)}$ and then for each type of inclusion i define auxiliary tensors

$$K^{(i)} \; = \; (C^{(0)} - (C^{(0)} - C^{(i)})M^{(i)})^{-1} \tag{8}$$
$$Q^{(i)} \; = \; C^{(0)}(I - M^{(i)}) \tag{9}$$

and the characteristic tensor for the medium

$$J = \left(I - \Sigma_i\phi_i K^{(i)}C^{(i)}C^{(0)^{-1}}Q^{(i)} \right)^{-1} \tag{10}$$

The elastic constant tensor for the composite medium is then given by

$$C^{\text{eff}} = \left(I + J\Sigma_i\phi_i K^{(i)}C^{(i)} \right) C^{(0)^{-1}} \tag{11}$$

The expression for thermal conductivity is similar. Note that, unlike the original treatment in [40], we associate a different tensor with each component. For simplicity, in the results presented below we have taken the reference properties $C^{(0)}$ to be those of the matrix

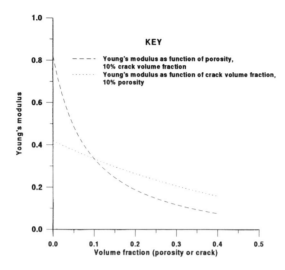

Figure 4: Dependence of Young's modulus Y on morphology for a porous coating: the two curves represent Y for a material with varying porosity and a constant 10 percent volume fraction of cracks, and for the same material with a 10 percent porosity and a varying volume fraction of cracks.

material and have not attempted a self-consistent solution. This gives a simple estimate of the effective properties rather than bounds. Since in this case the 'upper' and 'lower' bound curves cross each other, this avoids considerable complexity. The tensors A and B may be taken, for a cracked porous medium, to be those characteristic of spherical pores and highly oblate spheres to represent disk-like cracks.

Figure 4 shows the dependence of Young's modulus Y on morphology: the two curves represent Y for a material with varying porosity and a constant 10 percent volume fraction of cracks, and for the same material with a 10 percent porosity and a varying volume fraction of cracks. The differences arising from the different microstructures is clear.

CONCLUSIONS

The study of hetero-structures throws up a variety of related problems at the research community. It is not possible, and may not even be desirable, to attempt to model all the features of hetero-structures at the same level of detail. Many effects arise at characteristic length-scales and are best modelled on the scale where they appear. There is often a hierarchy involved. The meaning of the parameters of a given length-scale is defined in terms of smaller scales; the results are fed into larger-scale simulations. In this paper we have illustrated this by nucleation models defined in terms of atomic processes and bulk properties as determined by the distribution of cracks and pores.

ACKNOWLEDGEMENTS

The authors are grateful to CEC (Brite-Euram) programme and the Underlying Programme of UKAEA (Harwell) for funding on the work on plasma-sprayed coatings, and to numerous co-workers, in particular Professor John Venables (Sussex and Arizona) and Doctor Paul Chalker (Liverpool), for stimulating discussions.

REFERENCES

1. S.C. Jain, Germanium-Silicon Strained Layers and Heterostructures Academic Press, Boston 1994.

2. C. Priester and M. Lannoo, Current Opinion in Solid State and Mater. Sci. **2** p. 716 (1998).

3. A.E.M. de Veirman, J. Timmers, F.J.G. Hakkens, J.F.M. Cillessen and R.M. Wolf, Philips J. Res. **47** p. 185 (1993).

4. L. Dong, R.W. Smith and D.J. Srolovitz, J. Appl. Phys. **80** p. 5682 (1996).

5. S.C. Jain, T.J. Gosling, J.R. Willis, R. Bullough and P. Balk, Phil. Mag. A**65** p. 1151 (1992).

6. J.S. Speck and W. Pompe, J. Appl. Phys **76** p. 466 (1994); J.S. Speck, A. Seifert and W. Pompe, J. Appl. Phys **76** p. 477 (1994).

7. S.C. Jain, A.H. Harker and R.A. Cowley, Phil. Mag. A**75** p. 1461 (1997).

8. J.A. Venables, G.D.T. Spiller and M. Hanbücken M, Rep. Prog. Phys. **47** p. 399 (1984).

9. J.A. Venables, Phys. Rev. B **36** p. 4153 (1987).

10. J.H. Harding, J.A. Venables and A.M. Stoneham, Phys. Rev. B (in press) 1998.

11. J.A. Venables, Surf. Sci. **299-300** p. 798 (1994); for CaF_2 results see K.R. Heim, S.T. Coyle, G.G. Hembree, J.A. Venables and M.R. Scheinfein, J. Appl. Phys. **80** p. 1161 (1996).

12. A.D. Gates and J.L. Robins, Appl Surf Sci **48-9** p. 154 (1991).

13. H. von Harrach, Thin Solid Films **22** p. 30 (1974).

14. E.M. Chan, M.J. Buckingham and J.L. Robins, Surf. Sci. **67** p. 285 (1977).

15. D. Papajova, S. Nemeth, W.E. Hagston, H. Sitter and M. Veseley, Thin Solid Films, **267** p. 47 (1995).

16. T.X.T. Sayle, C.R.A. Catlow, D.C. Sayle, S.C. Parker and J.H. Harding, Phil. Mag. A**68** p. 565 (1993).

17. D.C. Sayle, S.C. Parker and J.H. Harding, J. Mater. Chem. **4** p. 1883 (1994).

18. A.P. Sutton and R.W. Balluffi Interfaces in Crystalline Solids Oxford U.P., Oxford, 1995.

19. M.W. Finnis, J. Phys. Cond. Mater. **8** p. 5811 (1996).

20. J.W. Matthews, D.C. Jackson and A. Chambers, Thin Solid Films **26** p. 129 (1975).

21. S.C. Jain, A.H. Harker, A. Atkinson and K. Pinardi, J. Appl. Phys. **78** p. 1630 (1995).

22. S.C. Jain, B. Dietrich, H. Richter, A. Atkinson and A.H. Harker, Phys. Rev. B**52** p. 6247 (1995).

23. C. Kisielowski, J. Kruger, S. Ruvimov, T. Suski, J.W. Ager, E. Jones, Z. Liliental-Weber, M. Rubin, E.R. Weber, M.D. Bremser and R.F. Davis, Phys. Rev. B**54** p. 17745 (1996).

24. A.L. Barabasi and H.E. Stanley, Fractal concepts in surface growth Cambridge U.P. Cambridge, 1995.

25. S. Cirolini, M. Marchese, G. Jacucci, J.H. Harding and P.A. Mulheran, Materials Design and Technology (ed. T.J. Kozik) **62** ASME 1994.

26. G.K. Batchelor, Ann. Rev. Fluid. Mech. **6** p. 227 (1991).

27. J.C. Maxwell Treatise on Electricity and Magnetism 2nd Edition, Clarendon Press 1881, Vol 1 p. 400.

28. L. Lorenz, Ann. Phys. Lpz. **11** p. 70 (1880).

29. Lord Rayleigh (J.W. Strutt), Phil. Mag. **34** p. 481 (1882).

30. D.A.G. Bruggeman, Ann. der Phys. **24** p. 636 (1935).

31. D. Polder and J.H. van Santen, Physica (Utrecht) **12** p. 257 (1946).

32. B. Budiansky, J. Mech. Phys. Sol. **13** p. 223 (1965).

33. T.T. Wu, Int. J Solids Struct. **2** p. 1 (1966).

34. J.D. Eshelby, Proc. Roy. Soc. A**252** p. 561 (1957).

35. T.W. Clyne and P.J. Withers, An Introduction to metal matrix composites Cambridge U.P., Cambridge, 1993.

36. T. Mura, Micromechanics of defects in solids Nijhoff, The Hague, 1987.

37. M. Ferrari, Compos. Eng. **4** p. 37 (1994).

38. R. Hill, J Mech. Phys. Sol. **11** p. 357 (1963).

39. Z. Hashin, J Mech. Phys. Sol. **36** p. 719 (1988).

40. O.B. Pedersen and P.J. Withers, Phil. Mag. A**65** p. 1217 (1992).

THE ROLE OF THE DIFFUSION MECHANISM IN MODELS OF THE EVOLUTION OF MICROSTRUCTURE DURING PHASE SEPARATION

T.T. Rautiainen and A.P. Sutton
Department of Materials, University of Oxford, Parks Road, Oxford OX1 3PH, U.K.

ABSTRACT

We have studied phase separation and subsequent coarsening of the microstructure in a two-dimensional square lattice using a stochastic Monte Carlo model and a deterministic mean field model. The differences and similarities between these approaches are discussed. We have found that a realistic diffusion mechanism through a vacancy motion in Monte Carlo simulations is cruicial in producing different coarsening mechanisms over a range of temperatures. This cannot be captured by the mean field model, in which the transformation is governed by the minimization of a free energy functional.

INTRODUCTION

When quenched from a disordered regime into a miscibility gap in its phase diagram, an alloy decomposes either spinodally or by a nucleation and growth mechanism. The dominant process observed is the separation of the single phase disordered system into a two phase mixture, and the subsequent coarsening of these phases. In the case of Ostwald ripening larger particles grow at the expense of smaller particles. If the minority phase occupies a negligible volume fraction, the rate of coarsening for droplets is proportional to the diffusivity of minority atoms within the majority phase, and the kinetics is expected to follow the Lifshitz-Slyozov-Wagner law [1]. Huse [2] considered some finite-time corrections to LSW-theory due to diffusion along domain boundaries. This extends the theory to cases, where the structure forms interconnected domains. At temperatures well below T_c the processes at particle interfaces start to dominate. Thermal energy is insufficient for supplying diffusing solute atoms into the matrix, and transport across interfaces becomes the limiting process. In this case the coarsening proceeds through Brownian motion of particles rather than LSW evaporation and condensation of atoms. Particles diffuse without changing size, and coarsening is a result of the coalescence of the diffusing particles [3,4].

Evolving microstructures have been modelled using mean field approaches and Monte Carlo simulations. Chen and Khachaturyan [5,6] used a microscopic mean field equation for the time evolution of single-site occupation probability functions. This model assumes that there exists a free energy functional describing both equilibrium and non-equilibrium states of a system, and that the time evolution of the local order parameter (in this case composition) is linear with respect to the local thermodynamic driving force. Vaks et al. [7,8] avoided this linearization and proposed a mean field equation to treat configurational kinetics at an arbitrary degree of non-equilibrium. However, in mean field models dynamics of fluctuations are not taken into account, and the true mechanism of atomic diffusion is not treated explicitly. These limitations can be circumvented in Monte Carlo simulations [9,10,11,12,13]. Most Monte Carlo studies on phase separation and coarsening kinetics have been done with direct exchange Kawasaki dynamics, although in real metals diffusion usually proceeds via vacancy jumps.

In this paper we have studied phase separation and subsequent coarsening using the mean field microscopic approach of [5,14] and Monte Carlo simulations with vacancy dynamics. Some comparisons with Kawasaki dynamics are also made.

DESCRIPTION OF MODELS

In the mean field approach we use a diffusion equation due to Khachaturyan [5,14], which is a counterpart of the continuum Cahn-Hilliard equation at the microscopic level:

$$\frac{dn(\mathbf{r}, t)}{dt} = \sum_{\mathbf{r}'} M(\mathbf{r} - \mathbf{r}') \frac{\delta F}{\delta n(\mathbf{r}', t)}. \tag{1}$$

$n(\mathbf{r}, t)$ is the probability that site $\mathbf{r}$ is occupied by an A atom, and $1 - n(\mathbf{r}, t)$ the probability that it is occupied by a B atom ($0 \leq n(\mathbf{r}, t) \leq 1$). $M(\mathbf{r} - \mathbf{r}')$ are probabilities for diffusional jumps between nearest neighbour lattice sites $\mathbf{r}$ and $\mathbf{r}'$, and in order to conserve the total number of atoms the condition $\sum_{\mathbf{r}} M(\mathbf{r}) = 0$ is required. Following Khachaturyan we assume that the values of M between nearest neighbour sites are constant. The functional derivative $\frac{\delta F}{\delta n(\mathbf{r}, t)}$ is the thermodynamic driving force for the transformation, and in equilibrium it becomes zero. The configurational entropy is treated in the Bragg-Williams approximation, and the free energy is then written as

$$F = \frac{1}{2} \sum_{\mathbf{r}, \mathbf{r}'} V(\mathbf{r} - \mathbf{r}') n(\mathbf{r}) n(\mathbf{r}') + k_B T \sum_{\mathbf{r}'} n(\mathbf{r}') \ln n(\mathbf{r}') + [1 - n(\mathbf{r}')] \ln[1 - n(\mathbf{r}')] \tag{2}$$

where $V(\mathbf{r} - \mathbf{r}') = \epsilon_{aa}(\mathbf{r} - \mathbf{r}') + \epsilon_{bb}(\mathbf{r} - \mathbf{r}') - 2\epsilon_{ab}(\mathbf{r} - \mathbf{r}')$ is the usual interaction parameter in quasi-chemical theory. The time evolution of the sample is obtained by taking a Fourier transformation of Eq. (1) and integrating it in reciprocal space:

$$\frac{d\tilde{n}(\mathbf{k}, t)}{dt} = \tilde{M}(\mathbf{k}) \left[\tilde{V}(\mathbf{k}) \tilde{n}(\mathbf{k}, t) + k_B T \left\{ \ln \left(\frac{n(\mathbf{r}, t)}{1 - n(\mathbf{r}, t)} \right) \right\}_{\mathbf{k}} \right]. \tag{3}$$

With interactions up to second nearest neighbours in a square lattice, the Fourier transform of the interaction energy $V(\mathbf{r})$ is $\tilde{V}(\mathbf{k}) = 2V_1 [\cos(2\pi h/L) + \cos(2\pi l/L)] + 4V_2 \cos(2\pi h/L) \cos(2\pi l/L)$ where $\mathbf{k} = \frac{2\pi}{L}(h, l)$, $h, l = 0, \pm 1, ..., \pm L/2$ is the wave vector in the first Brillouin zone, and L is the system dimension. V_1 and V_2 are the interaction parameters between the first and second nearest neighbours. Allowing only nearest neighbour jumps we have $\tilde{M}(\mathbf{k}) = -4L_1 [\sin^2(\pi h/L) + \sin^2(\pi l/L)]$. Equation (3) is solved by numerical differencing on a mesh of $\mathbf{k}$-points, the number of which is equal to the number of lattice sites in the real space computational cell. Time integration is done using a Runge-Kutta algorithm.

We use a 128×128 lattice with periodic boundary conditions. Initially each lattice site is assigned a composition, which deviates by a small random amount from an average value, corresponding to a high temperature disordered state. Atomic interactions between nearest and next nearest neighbours are chosen to be $V_1 = -1$ and $V_2 = -1/\sqrt{2}$, in arbitrary units. The mean field critical temperature for phase separation is $k_B T_c^{MF} = |V_1 + V_2|$.

The second approach employed in this study is a Monte Carlo model with a second order residence time algorithm. One lattice site is vacant, corresponding to a vacancy concentration of $1/128^2 = 6.1 \times 10^{-5}$. The total energy of the system can be expressed as an Ising type Hamiltonian

$$H = J_1 \sum_{<NN>} \sigma_i \sigma_j + J_2 \sum_{<NNN>} \sigma_i \sigma_j + C_1 \tag{4}$$

in which $J_1 = \frac{1}{4}(\epsilon_{aa}^{(1)} + \epsilon_{bb}^{(1)} - 2\epsilon_{ab}^{(1)})$ and $J_2 = \frac{1}{4}(\epsilon_{aa}^{(2)} + \epsilon_{bb}^{(2)} - 2\epsilon_{ab}^{(2)})$ are the NN and NNN interactions and C_1 is a constant. Interactions between the vacancy and atoms are zero. The parameters are made comparable to the mean field model by choosing them negative, and $J_1/J_2 = \sqrt{2}$. By using grand canonical ensembles and a fourth order cumulant method [15] the critical temperature was estimated to be $k_B T_c = 2.58|J_1 + J_2|$. The initial state for the simulation is a completely disordered state, where atomic occupations are determined by a random number generator.

All changes in site occupancies are effected by the motion of the single vacancy, which performs jumps only to nearest neighbour positions. Each neighbouring atom i is vibrating with frequency ν_i. In order to jump to the vacant lattice site, an atom has to overcome an activation energy barrier of height $E_a^i = E_s^i - E_1^i$, where E_s^i is the environmentally dependent saddle point energy and E_1^i is the initial energy state of the atom i. Activation energy barriers for vacancy migration are sensitive to the local atomic environment. In our model this is reflected by making the saddle point energy a function of both the initial and final states: $E_s^i = \frac{E_1^i + E_2^i}{2} + e_0$. The constant e_0 is fixed and chosen such that $E_s^i - E_1^i$ is always positive. The effect of the final state is included directly through the saddle point energy: a higher E_2 increases the saddle point (and activation) energy for the jump. For the ith atom surrounding the vacancy the transition probability per unit time is $\omega_i = \nu_i exp(-E_a^i/k_B T)$, ν_i being the attempt frequency. The relative probabilities for vacancy jumps are

$$\alpha_i = \frac{\nu_i e^{-E_a^i/k_B T}}{\sum_j \nu_j e^{-E_a^j/k_B T}}. \tag{5}$$

In contrast to the usual Metropolis algorithm, a vacancy jump takes place at every step, but the time increment associated with the step is no longer constant [16,17,18] A time scale is defined through the atomic vibration frequencies and activation barrier heights. The mean residence time τ is the average time the system is expected to remain in a given configuration

$$\tau = \frac{1}{\sum_i \nu_i e^{-E_a^i/k_B T}} \tag{6}$$

where the sum is over the nearest neighbours of a vacancy.

The first order residence time algorithm does not take into account the possibility that the vacancy may jump straight back to its previous lattice position. This corresponds to a rejection in a sense that the atomic configuration has not changed, but time has elapsed. In order to eliminate direct reversals, the transition probabilities should be modified, and the second order residence time includes the contribution of reversal jumps correspondingly. We have used the second order residence time algorithm in this work, and we refer the reader to a full derivation of this in Ref. [19].

RESULTS

The critical temperatures for the mean field and Monte Carlo models are different, but the simulations are run for the same normalized temperatures T/T_c. Fig. 1 shows a mean field quench to $T/T_c = 0.16$ with $c = 0.15$. We see small droplets, that have a spherical shape, which is maintained throughout coarsening. Small particles gradually disappear, which can be understood as evaporation at the expense of the larger nearby particles. No particle diffusion and coagulation was observed. In their recent Monte Carlo study, Fratzl and Penrose [13] found that at low temperatures the particles move as a whole rather than evaporate, which is supported by Binder-Stauffer phenomenological theory [4]. Our Monte Carlo simulations suggest similar behaviour. At temperatures close to T_c in Fig. 2.(a) the evaporation mechanism dominates. The particles are rough and there are isolated solute atoms in the matrix. At $T/T_c = 0.50$ in Fig. 2.(b) the LSW-evaporation is still active, but the particle diffusion and coagulation mechanism is taking over. At low temperatures the vacancy prefers to stay at interfaces, where it reduces the number of energetically expensive A-B bonds. The center of mass of the particles moves due to the movement of surface atoms of the particles, or when the vacancy enters the particle. Occasionally

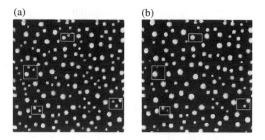

(a)　　　　　　　(b)

Figure 1: Snapshots of evolving microstructures obtained with the mean field model with $c = 0.15$ and $T/T_c = 0.16$. (b) is a later time than (a). Boxes indicate some evaporation events.

the vacancy escapes and attaches itself to another particle, which then undergoes Brownian motion. Coarsening is a result of coagulation of these diffusing particles (Fig. 2.(c)). In Monte Carlo vacancy simulations at low temperatures the particle interfaces are alinged along $\langle 11 \rangle$ directions. The anisotropy is a result of the fact that the vacancy is a localized defect that moves easily along $\langle 10 \rangle$ facets and with greater difficulty along $\langle 11 \rangle$ facets [20]. With Kawasaki dynamics $\langle 10 \rangle$ facets were favoured instead.

In order to confirm the Brownian motion at low temperatures ($T/T_c < 0.5$), further Monte Carlo simulations were carried out. Initially two particles of sizes 13 and 113 atoms were placed at sites (42,42) and (85,85) in our system, and the movement of their centers of masses was monitored. As can be seen in Fig. 3.(a), in Monte Carlo simulations the particles move randomly.

(a)　　　　　　(b)　　　　　　(c)

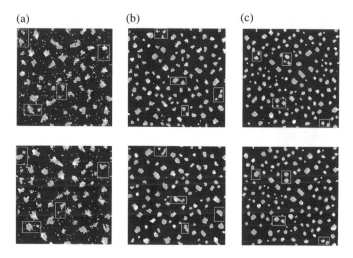

Figure 2: Snapshots of evolving microstructures with $c = 0.15$ obtained with the Monte Carlo model: (a) $T/T_c = 0.93$, (b) $T/T_c = 0.50$ and (c) $T/T_c = 0.16$. The bottom row is later time than the top row. Boxes indicate some coagulation and evaporation events.

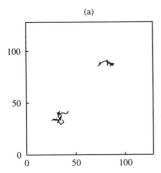
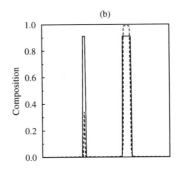

Figure 3: Time evolution of two particles at $T/T_c = 0.25$. (a) A Monte Carlo result: trajectories of the centers of masses of the particles. (b) A mean field result: composition profiles through the sample showing the shrinkage of the smaller particle, but no Brownian motion. Solid line shows the initial composition.

This Brownian motion was not so clear with Kawasaki dynamics, and as shown in Ref. [13], the LSW-evaporation dominates even at low temperatures. In the mean field model the initial confinement of all the solute to the particles is not an equilibrium distribution owing to the configurational entropy term in the free energy functional, Eq. (2). Some solute dissolves in the matrix so that the concentration in the particles is less than one. We set the total amount of solute equal to that in the Monte Carlo simulation. The mean field result in Fig. 3.(b), indicates no Brownian motion, but the smaller particle gradually evaporates. Changing the temperature did not change this qualitative picture. When the overall concentration was increased, no evaporation was observed, and particles grew by absorbing solute atoms from the matrix. When the particles were initially brought close together, their centers of masses moved away from each other, because the region between the particles became depleted of solute atoms, and growth could occur only in directions away from their joint center of mass.

SUMMARY

In this paper we have demonstrated the influence of a realistic diffusion mechanism on descriptions of phase separation and coarsening. The main difference between the models used is the way in which thermal equilibrium is approached, either in a deterministic manner by minimizing a free energy functional in the mean field model or in a stochastic manner through vacancy jumps in the Monte Carlo model. The Monte Carlo model produced different coarsening mechanisms: at low temperatures particle diffusion and coagulation while at high temperatures solute evaporation and condensation. The presence of a vacancy also affected the particle morphology at low temperatures. By contrast, the mean field model did not show any tendency to faceting, and it failed to produce particle diffusion at low temperatures, with LSW-evaporation dominating at all temperatures.

ACKNOWLEDGMENTS

TTR gratefully acknowledges the financial support from the Osk. Huttunen Foundation, Finland. These simulations were carried out in the Materials Modelling Laboratory of the Department of Materials, Oxford University.

REFERENCES

1. I.M. Lifshitz, V.V. Slyozov, J. Phys. Chem. Solids **19**, 35 (1961).

2. D.A. Huse, Phys. Rev. B **34**, 7845 (1986).

3. K. Binder, D. Stauffer, Phys. Rev. Lett. **33**, 1006 (1974).

4. J.D. Gunton, M. San Miguel, P. Sahni, in *Phase Transitions and Critical Phenomena*, ed. by C. Domb and J.H. Lebowitz, (Academic, London, 1983), Vol. 8, pp. 269-466.

5. Y. Wang, L.Q. Chen, A.G. Khachaturyan, in *Computer Simulation in Materials Science*, ed. by H.O. Kirchner *et al.*, (Kluwer, Netherlands, 1996), pp. 325-371.

6. L.Q. Chen, A.G. Khachaturyan, Acta Metall. Mater. **39**, 2533 (1991).

7. V.G. Vaks, S.V. Beiden, V. Yu. Dobretsov, JETP Lett. **61**, 68 (1995).

8. V. Yu. Dobretsov, V.G. Vaks, G. Martin, Phys. Rev. B **54**, 3227 (1996).

9. K. Yaldram, K. Binder, Acta Metall. Mater. **39**, 707 (1991).

10. P. Fratzl, O. Penrose, Phys. Rev. B **50**, 3477 (1994).

11. C. Frontera, E. Vives, T. Castán, A. Planes, Phys. Rev. B **53**, 2886 (1996).

12. J.G. Amar, F.E. Sullivan, R.D. Mountain, Phys. Rev. B **37** 196 (1988).

13. P. Fratzl, O. Penrose, Phys. Rev. B **55**, R6101 (1997).

14. A.G. Khachaturyan, *Theory of Structural Transformations in Solids* (Wiley, New York, 1983).

15. K. Binder, D.W. Heerman, *Monte Carlo Simulation in Statistical Physics*, (Springer-Verlaag, Berlin, 1988).

16. A.B. Bortz, M.H. Kalos, J.L. Lebowitz, J. Comp. Phys. **17**,10 (1975).

17. M.A. Novotny, Computers in Physics **9**, 46 (1995).

18. W.M. Young, E.W. Elcock, Proc. Phys. Soc. **89**, 735 (1966).

19. M. Athènes, P. Bellon, G. Martin, Phil. Mag. A **76**, 565 (1997).

20. T.T. Rautiainen, A.P. Sutton, to be published

SIMULATING PHASE TRANSFORMATIONS WITH THE CAHN-HILLIARD EQUATION
- POTENTIAL AND LIMITATIONS -

Lothar Löchte, Günter Gottstein
Institut für Metallkunde und Metallphysik, RWTH Aachen, Germany
Kopernikusstr. 14, 52074 Aachen, loechte@imm.rwth-aachen.de
Collaborative Research Center 370 „Integral Modeling of Materials" of the DFG

ABSTRACT

An extension of the classical Cahn-Hilliard equation, including elastic interactions is presented. This generalized diffusion equation allows real time simulation of phase-transformations, such as GP-zone formation in a near-commercial AlCu alloy. No a priori assumptions about kinetics, diffusion fields, as well as precipitate shape are necessary. Shape of precipitates and kinetic of simulated GP-zone formation are in qualitatively good agreement with experiments of Sato and Takahashi, and the concurrenltly derived interfacial energies are reasonable.

Two algorithms for a numerical solution of the Cahn-Hilliard equation are presented. While the simple finite difference scheme does not converge, a slightly more complicated Fourier spectral method behaves reasonable as tested for a synthetic chemical potential.

INTRODUCTION

The mechanical properties of advanced commercial alloys are commonly due to specific but metastable microstructures, which are most often generated by diffusion controlled phase transformations. The wide range of such phase transitions can be described by diffusion equations, which have to account concentration gradients (interfaces), as well as elastic effects, such as volume misfit of precipitates. The master equation for such kinetics is based on the classical Cahn-Hillard equation (CHE) and its application to the simulation of phase transformations of near-commercial alloys, such as AlCu, will be discussed in this paper. A short term goal of the underlying project is to simulate the kinetics of GP-zone formation in age-hardenable AlCu alloys and to fit the interfacial energy to the evolving shape of the GP-zones. Eventually it is planned to describe the sequential development of the different metastable phases (GP-zones, Θ' and Θ) occurring in this class of alloys. Within the framework of the Collaborative Research Center 370 it is intended to connect these results to simulations of deformation and recrystallization.

EXTENDED CAHN-HILLIARD MODEL

In the extended Cahn-Hilliard model [1, 2, 3, 4] we formulate the total free energy F, with the free energy density f, of the alloy system to consist of three parts, the chemical $F_{chemical}$, the gradient/interfacial $F_{gradient}$ and the elastic part $F_{elastic}$. The first part characterizes a driving force, while the second and third part characterize a backdriving force.

$$F = \int f \, dV = \int [f_{chemical} + f_{gradient} + f_{elastic}] \, dV \qquad (1)$$

Mat. Res. Soc. Symp. Proc. Vol. 529 © 1998 Materials Research Society

The total thermodynamic driving force of the phase transformation is the variational derivative, $\delta F/\delta c$, of the total free energy F, also known as the chemical potential μ. Minimizing the total free energy of the different phases in an alloy system gives the compositions of the phases, as well as their volume fraction in the alloy. For prediction of the phase transformation kinetics, we need the temporal and spatial evolution of each individual component.

$$\frac{\partial c}{\partial t} = \nabla\left(\frac{D}{RT} \cdot c(1-c)\nabla\left(\frac{\delta F}{\delta c}\right)\right) \tag{2}$$

For dilute systems we can simplify equation (2) to equation (3), the CHE,

$$\frac{\partial c}{\partial t} = \frac{D}{RT} \cdot c_0 \Delta\left(\frac{\delta F}{\delta c}\right) \tag{3}$$

with the following terms of the total free energy F (**Fig. 1, 2**):

$$F_{chemical} = \int[(1-c)\cdot\mu_{Al}^0 + c\cdot\mu_{Cu}^0 + RT(c\cdot\ln(c) + (1-c)\cdot\ln(1-c))$$
$$+ c(1-c)\sum_{j=0}^{2} L^j \cdot (1-2c)^j]\,dV \tag{4}$$

$$F_{gradient} = \int[\frac{\gamma}{2}|\nabla c|^2]\,dV \tag{5}$$

$$F_{elastic} = \frac{1}{2}\int\frac{dk^3}{(2\pi)^3}\tilde{A}(k,c_{ij},\Delta a)\cdot\tilde{c}^2 \tag{6}$$

$$\tilde{A}(k,c_{ij},\Delta a/a) = -\frac{(\Delta a/a\cdot(c_{11}+c_{12}))^2}{c_{11}}$$
$$\cdot\frac{(c_{11}-c_{12})(c_{11}-c_{12}-2c_{44})k_1^2k_2^2}{c_{11}c_{44}(k_1^2+k_2^2)^2 + (c_{11}+c_{12})(c_{11}-c_{12}-2c_{44})k_1^2k_2^2} \tag{7}$$

where μ_{Al}=-12.55kJ/mol, μ_{Cu}=-14.60kJ/mol, L^0=9.327kJ/mol, L^1=0.635kJ/mol, L^2=6.889kJ/mol) at T=423K [7], with the energy gradient coefficient γ, with c = c_{Cu}, the interaction coefficient A in two dimensions [3, 4, 5], with the wave vector k=(k_1,k_2) and the elastic constants c_{ij}.

In real space equation (6) corresponds to a double convolution and the elastic strain decreases with 1/distance^2 in two dimensions and with 1/distance^3 in three dimensions. It should be noted that this, as well as the different discretizations of the gradient/laplacian operator in the gradient part results in different backdriving forces for the phase transformation in two and three dimensions.

The extended Cahn-Hilliard equation then reads:

$$\frac{\partial c}{\partial t} = \frac{D}{RT} \cdot c_0 \cdot \Delta\left(\mu(c) - \gamma\cdot\Delta c + \int\tilde{A}\cdot\tilde{c}\cdot e^{ikr}\,dk^3\right) \tag{8}$$

For the diffusion coefficient D we assume an Arrhenius behaviour with an activation energy Q=135.34 kJ/mol and a preexponential factor $D_0=6.47 \cdot 10^{-5}$ m^2/s [5].
Following Cahn & Hilliard [2], the energy gradient coefficient γ is coupled to the interfacial energy σ by:

$$\sigma = \int_{c_\alpha}^{c_\beta} (2 \cdot \gamma \cdot \Delta f_{chemical})^{\frac{1}{2}} \, dc \qquad (9)$$

, where c_α and c_β are the concentrations within the two phases.

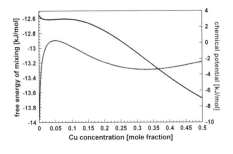
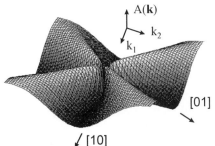

Fig. 1: *Free energy of mixing for GP-zones in* **Fig. 2:** *Elastic interaction coefficient A(k)*
AlCu at T=423K   *(Δa/a=10%,* c_{11}=108GPa, C_{12}=62GPa,
                                        c_{44}=28.3GPa)

NUMERICAL ASPECTS

For a numerical solution of the CHE two different methods were used, a simple Euler finite difference scheme and a Fourier spectral method. In case of a finite difference approach a forward finite difference scheme replaces the derivative in time (left-hand side of equation (8))

$$\frac{\partial c(\vec{r},t)}{\partial t} \rightarrow \frac{c^{n+1}(\vec{r}) - c^n(\vec{r})}{\tau^n} \qquad (10)$$

and the laplacian operator, reads, for sake of simplicity shown in one dimension

$$\Delta \alpha(c) \rightarrow \frac{\alpha(c_{i-1}^n) + \alpha(c_{i+1}^n) - 2\alpha(c_i^n)}{\Delta x^2} \qquad (11)$$

with the time step τ^n and the spatial discretization Δx.
A special time step control was implemented, to obtain the maximum possible decrease of the total free energy in each integration step.
The partial differential equation can be transformed into a sequence of ordinary differential equations in Fourier space, which can be easily solved with a finite difference scheme with respect to $\partial c / \partial t$ [8]. For the right hand side of equation (12) a semi-implicit discretization was used:

73

$$\frac{\partial \tilde{c}(\mathbf{k})}{\partial t} = -\frac{D}{RT} \cdot c_0 \cdot |\mathbf{k}|^2 \cdot \left\{ \left(\frac{\delta F}{\delta c} \right) \right\}_{\mathbf{k}} \qquad \text{, with} \quad |\mathbf{k}|^2 = k_x^2 + k_y^2 \qquad (12)$$

and with $\qquad \tilde{c}(\mathbf{k}) = \int c(\mathbf{r}) \cdot e^{-i\mathbf{k}\mathbf{r}} d\mathbf{r}^3 \qquad (13)$

The concentration, as well as the variational derivative of the free energy are Fourier transformed in each integration step by use of a FFT algorithm [9].

SIMULATION RESULTS

The 2D-simulations (**Fig. 3**) start with a supersaturated solid solution of uniform distribution of 5at.% Cu, quenched in from solutionizing temperature and subjected to thermal fluctuations. **Fig. 3b** shows a simulation without elastic interactions, while in **Fig. 3c** a misfit of $\Delta a/a$=10%, which is typical for GP-zones, is assumed.

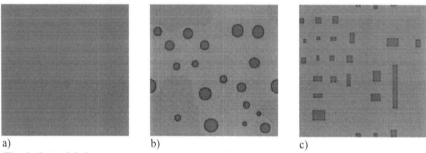

a)   b)   c)

Fig. 3: Spinodal decomposition, c_0=5at.%, T=423K, NxN=256x256, numerical method: finite difference scheme a) t=0s. solid solution (quenched) b) $\Delta a/a$=0%, σ=4.5·10^{-3}J/m^2, b) $\Delta a/a$=10%, σ=2·10^{-3}J/m^2

With circular precipitates for $\Delta a/a$=0% and precipitates growing along the [10] and the [01] directions for $\Delta a/a$=10% both simulations qualitatively give correct results. Although there is little information about the interfacial energy of GP-zones in <u>Al</u>Cu in the literature, we feel that the given values are physically meaningful.

Next we start with a nucleus and an average Cu concentration of 2at.% in the matrix, $\Delta a/a$=10%, and simulate the nucleus growth for two spatial grid sizes, i.e. Δx=0.1nm (**Fig. 4b**) and the same simulation with Δx=0.05nm (**Fig. 4c**).
Due to an insufficient few number of grid points within the interface (**Fig. 4c**), this algorithm (finite difference scheme) does not converge. Nevertheless the interfacial energy of σ=2·mJ/m^2 is physically meaningful, even if it is not the quantitative correct value. The evolution of the GP-zone size is in good agreement with experiments of Sato and Takahashi [10]. The remaining discrepancy can be attributed to the absence of coarsening and the assumption of concentration independent elastic constants.

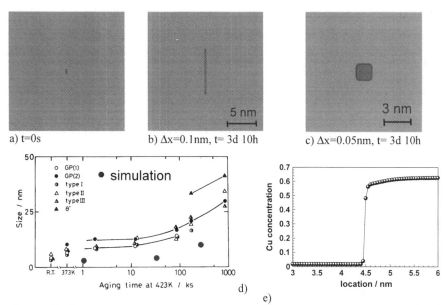

a) t=0s   b) Δx=0.1nm, t= 3d 10h   c) Δx=0.05nm, t= 3d 10h

d)

e)

Fig. 4: *Nucleus growth, at T=423K, for t=3d 10h, with $\Delta a/a$=10% and σ=2·10^{-3}J/m^2, numerical method: finite difference scheme a) solid solution b) Δx=0.1nm c) Δx=0.05nm, d) comparison with experiments of Sato and Takahashi [10] e) cut along the longitudinal direction of the GP-zone of b)*

In order to eliminate the divergence problem, we implemented the Fourier spectral method (FSM) and tested it with an artificial chemical potential, again for two spatial resolutions (**Fig. 5**). Obviously, there is now a better convergence of the numerical solution of the Cahn-Hilliard equation. In contrast to the simulation with the finite difference scheme, the FSM yields elongated precipitates with only slightly different aspect ratios, independent of the spatial grid size. Also, the largest time step to be chosen for the FSM is roughly one magnitude higher than for the common finite difference scheme.

Next we plan to use the FSM for the simulation of GP-zone formation in the near commercial AlCu alloy.

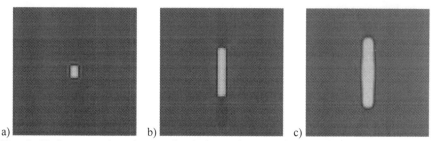

a)   b)   c)

Fig. 5: *Nucleus growth with a artificial chemical potential a) single nucleus and super-saturated matrix b) t=10, Δx=1, c) t=10, Δx=0.5*

CONCLUSIONS

By use of thermodynamic data the extended Cahn-Hilliard equation predicts the correct microstructure of a near-commercial AlCu alloy. The kinetics of GP-zone formation, as well as the shape and the crystallographic orientation are qualitatively in good agreement with experimental observations. No a priori assumptions concerning the shape of precipitates are necessary. The kinetics of the phase transformation are dominated by the mobility M of the diffusiong atoms and yield the real-time kinetics without further assumptions. In case of insufficiently few points within the interface a simple finite difference scheme does not converge and needs to be replaced by a numerically more stable algorithm. By use of a Fourier spectral method, proposed by Chen and Shen a fast and stable solution can be obtained.

REFERENCES

[1]   J. W. Cahn; Acta Metall.; 10 (1962) 179-183.

[2]   J.W. Cahn, J. E. Hilliard; J. Chem. Phys.; 28 (1958) 258-267.

[3]   A.G. Khachaturyan; Theory of structural transformation; John Wiley & Sons, New York, Chichester, Brisbane, Toronto, Singapore, 1983

[4]   Y. Wang, L.-Q. Chen, A.G. Khachaturyan; "Computer Simulation in Materials Science", eds. H.O. Kirchner et. al., Kluwer Academic Publishers (1996) 325-371.

[5]   L. Löchte, A. Gitt, G. Gottstein, to be published

[6]   „Smithells Metals reference book", ed. By A.E. Brandes, Butterworths, London, Boston, 1983

[7]   I. Hurtado, B. Meurer, L. Löchte, J. Dünnwald, P.J. Spencer, D. Neuschütz, Proc. 10th Congress of the Int. Fed. For Heat Treatment and Surface Eng., Grighton, England, Sept. 1996, to be published in 1998

[8]   L.Q. Chen, J. Shen; Applications of Semi-Implicit Fourier-Spectral Method to Phase-Field Equations, Comp. Phys. Comm., Feb. 1998

[9]   M. Frigo, S.G. Johnson; FFTW: An Adaptive Software Architecture for the FFT; to be published at Int. Conf. On Acoustics, Speech and Signal Processing, May 1998, Seattle, Washington

and http://theory.lcs.mit.edu/~fftw/

[10]   T. Sato, T. Takahashi; Scripta Metall.; 22 (1988), 941-946

Potts Model Simulation of Grain Size Distributions during Final Stage Sintering

P. Zeng*, V. Tikare**
*Center for Materials Simulations, IMS, University of Connecticut, Storrs, CT 06269
**Sandia National Lab, Computational Materials Modeling, Albuquerque, NM 87185-1411

ABSTRACT

The Potts Monte Carlo model was used to simulate microstructural evolution and characterize grain size distribution during the final stages of sintering. Simultaneous grain growth, pore migration and pore shrinkage were simulated in a system with an initial porosity of 10% with varying ratios of grain boundary mobility to pore shrinkage rates. This investigation shows that the presence of pores changes the grain size distribution and the topological characteristics due to pinning of grains by pores. As pores shrink away, their pinning effect decreases. Once pore shrinkage is complete, normal grain growth is achieved.

INTRODUCTION

During the final stages of sintering, pore channels along grain boundaries begin to shrink and pores are isolated on grain boundaries and triple junctions then shrink continuously and may disappear altogether[1]. However, in many cases, pores may break away from grain boundaries and become trapped within grains, resulting in some amount of residual porosity[2]. These observations neglect the importance of surface diffuison and bulk diffusion during final stage sintering. For pore shrinkage in the final stage of sintering, grain boundary diffusion is the most important mechanism. However, surface diffusion is responsible for pore mobility and will affect both sintering microstructure and sintering kinetics.

In this paper, we have developed a modified Potts Monte Carlo simulation algorithm to study final stage sintering, which is based on the previous development on the microstructure evolution such as grain growth and pore migration[3]. Multiple kinetics are incorporated by assigning different Monte Carlo probability to different mechanisms based on the experimental conditions which exist during final stage sintering. Using a 2D Potts model, microstructural evolution, topological distribution and grain size distribution are discussed in detail.

MODEL AND SIMULATION METHOD

The two-dimensional Potts model was used to study simultaneous grain growth, pore migration and pore shrinkage simulation. Grain structure is mapped onto a square lattice with periodic boundary conditions in both the X- and Y-directions. Each lattice is assigned a spin between 1 and q (q=100) which represents the different orientation of the grain in which the site is embedded. Each pore site is assigned a spin = -1. Each grain or pore site may be considered as a discrete domain, consisting of some billions of atoms, which maps to a 400×400 lattice to form a meso-scale microstructure. In this investigation, we chose the grain boundary energy and surface energy to be isotropic and independent of the grain orientation. Thus, we assign the bond energy, E_0, to be constant with direction and spin. First and second nearest neighbor interactions are considered as this results in lower anisotropy[4]. The equation of state for this system, also know as the Hamiltonian, was defined as

Mat. Res. Soc. Symp. Proc. Vol. 529 © 1998 Materials Research Society

$$E = \frac{1}{2}E_0 \sum_{i=1}^{N} \sum_{j=1}^{8}\left(1-\delta\left(S_i,S_j\right)\right) \qquad \text{eq.1}$$

where E_0 is the bond energy between neighboring sites of unlike spin ($s_i \neq s_j$, e.g. grain boundary or surface). The Hamiltonian counts the number of unlike bonds between all sites i and their 8 1st and 2nd nearest neighbors j.

Grain growth is simulated using Potts model: a grain site is chosen at random from the simulation space, then a new trial spin is chosen at random from q spins. The energy change is evaluated using eq.1. The change in energy for the grain growth step, ΔE, is then used to calculate the transition probability, P, using Boltzmann statistics as

$$P = \begin{cases} \exp(-\frac{\Delta E}{K_B T}) & \text{for } \Delta E > 0 \\ 1 & \text{for } \Delta E \leq 0 \end{cases} \qquad \text{eq.2}$$

where K_B is the Boltzmann constant and T is absolute temperature. The Metropolis algorithm[5] is used to determine if an exchange is accepted or not by choosing a random number between 0 and 1. If the random number is less than or equal to P, then the transition is accepted. If not, the transition is rejected. A Monte Carlo temperature T = 0 is used for grain growth simulation to eliminate the thermal fluctuations and has been shown to simulate grain growth well[3,6].

Pore migration is simulated using conserved dynamics. A pore site is picked at random, then a neighboring grain site is chosen also at random. A trial exchange of the grain site and the pore site with the grain site assuming the spin which results in the minimum energy is considered. The energy change of this trial exchange is calculated using eq.1. The transition probability is determined by applying eq.2 and the Metropolis algorithm is used to accept or reject the pore migration step. In this simulation, almost all pore migration events happen by grain sites moving along the pore-grain interface, thus, simulating pore migration by surface diffusion as shown by Tikare and Holm[3]. The mobility ratio of pore boundaries to grain boundaries was chosen to be 1:1 based on earlier work. Monte Carlo temperature of $K_B T = 0.5$ was chosen for pore migration also based on earlier work[3].

Pore shrinkage is assumed to occur by grain boundary diffusion only. First, we select a pore site at random. If the pore site is at a grain boundary, pore shrinkage is attempted. If it is an internal pore site then pore shrinkage is not permitted, as it is an intragranular pore. Pore shrinkage is simulated by replacing that pore site with a grain site which has a spin resulting in the minimum energy as calculated by eq. 1. We calculate the change in energy for shrinkage using eq. 1. The transition probability corresponding to this change is calculated using eq. 2 and finally, the Metropolis algorithm is used to accept or reject the transition.

The grain boundary mobility to pore migration ratio was held constant at 1 for all the simulations in this investigation. The pore migration to pore shrinkage ratio, m, was varied from 1,000 to 10,000. This was done by attempting m pore migration events for each pore shrinkage event. Pore migration and shrinkage are simulated under temperature $k_B T = 0.5$, this high temperature is necessary to sample higher entropy states required for simulation of pore migration and shrinkage. At lower temperature, pores would not have sufficient energy to diffuse in the microstructure[7].

Time in the Potts model is measured in units of Monte Carlo steps, MCS. At 1 MCS, the number of attempted changes is equal to the total number of lattice sites in the simulation. Simulations were run up to 10^5 MCS. Data was collected from four independent runs for each set of simulations run under identical conditions to obtain good statistics for grain size distributions and topologies.

RESULTS AND DISCUSSION

Previous study of pore migration and grain growth with no pore shrinkage showed that pores grow by coalescence and the grain growth is pinned by pores. In this simulation, we choose initial porosity of 10%, which is considered appropriate for the final stage sintering[8,9]. The starting microstructure for pore shrinkage simulations is obtained by only allowing grain growth and pore migration for a system with constant porosity of 10%. After the average grain size, measured in grain area, reaches 100 sites, pore shrinkage, along with grain growth and pore migration, is allowed to be active. This initial microstructure, Fig. 1a, shows that pores are present at grain triple junctions and almost all the triple junctions are occupied by pores. The pores and grains are equi-axed with smooth, regular interfaces.

A series of simulations with pore migration to pore shrinkage ratios of 1000, 3000, 7000 and 10,000 were run. These ratios were chosen to simulate the surface to grain boundary diffusion ratios typically observed in real materials systems.

For a two dimensional system with constant volume fraction of pores, pore growth rate was predicted by Tikare and Holm to be

$$R \propto t^{0.2} \qquad \text{eq.3}$$

where R is the grain radius and t is time. They obtained this relationship for the case where pores migrate through the microstructure by random walk and grow by coalescence (two or more pores become one large pore when they touch each other). Fig.2 shows that the pore growth curves, pore size vs. time, for systems with (a) grain growth and pore migration and (b-e) grain growth, pore migration and shrinkage at different pore migration to pore shrinkage ratios. For the system with no pore shrinkage, only grain growth and pore migration, pore growth curve shows that the pore growth exponent is about 0.15, which is a good agreement with the predicted pore growth exponent, indicating that pores do grow by coalescence. In the simulation with both pore shrinkage and pore migration, two competing events occur, pore growth by migrating pores which coalesce and pore shrinkage. The smaller pores disappear faster because of their higher curvature, thus increasing the mean distance between pores. This in turn makes pore coalesces via random walk less frequent. The competition between pore growth via coalescence and pore shrinkage via grain boundary diffusion tends to favor shrinkage as surface to grain boundary diffusion rate decreases as shown in Fig.2. However, at the slow shrinkage ratio, the pore coalescence could be seen in the microstructures shown in Fig. 1b to Fig. 1d. Pore breakaway could also been seen in those microstructures as pore shrink and no longer pin grain boundaries.

Fig.3 shows the grain growth curves for (a) grain growth without pores, (b) constant porosity, grain growth and pore migration and (c-f) different pore migration to shrinkage ratios. The grain growth kinetics display power law behavior with a grain growth exponent n, $R \propto t^n$, with $n = 0.5$[10]. In the simulations, the grain growth exponent for normal growth was $n = 0.5$, which is in good agreement with that predicted by the theory of single phase grain growth ($n_G = 0.5$[11]). For grain growth together with pore migration, grain growth is pinned by pores and thus, scales with pore growth with exponent $n_P = 0.2$ as indicated by Eq.3. The grain growth with pore migration simulation shows that $n = 0.25$ as shown in Fig.3, which is a reasonably good agreement. Grain growth with pore migration and shrinkage at different ratios shows that the grain growth is not log-linear. However, the grain growth curves for these simulations fall between those of normal grain growth and grain growth with pore migration. Furthermore, the simulation with highest pore

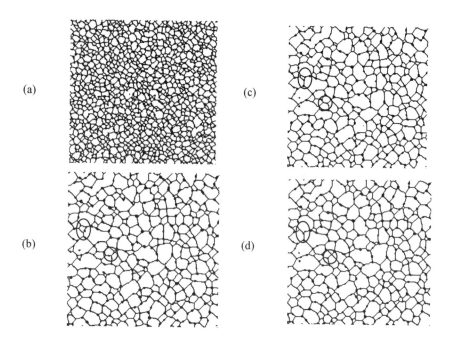

Fig.1 (a) Starting microstructure for pore shrinkage with initial porosity 10%, microstructure at (b) 80,000 (c) 90,000 and (d) 100,000 MCS at pore migration to pore shrinkage ratio m = 10,000. Pore coalescence events are circled.

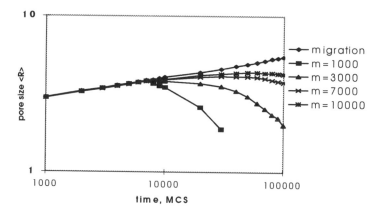

Fig.2 Pore size vs. time for different pore migration to pore shrinkage ratios m.

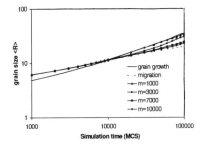 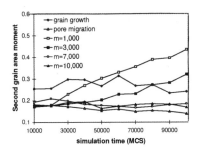

Fig.3 Grain growth curves with different pore migration to pore shrinkage rations m.

Fig.4 Second moment of grain area at different time for different pore migration to pore shrinkage rations m. For m=1,000, pores disappear at 40,000 MCS.

shrinkage rate is closest to the normal grain growth curve with the low shrinkage rates progressively deviating from the normal grain growth curve. These results indicate that grain growth is pinned by pores in all cases however, the pinning effect is decreasing as the pores shrink and disappear.

Fig.4 is the plot of the second moment of grain area as function of simulation time for all simulations. The second moment is the variance of the grain size distribution and is calculated as

$$\mu_2(A) = \frac{1}{n-1} \sum_{i=1}^{n} \frac{A_i^2}{<A>^2} - 1 \qquad \text{eq. 4}$$

where n is the number of grains, A is the grain area and <A> is the average area. Fig.4 shows that the systems with faster pore shrinkage rates tend to have broader grain size distributions than the normal grain growth size distribution. This effect not only persists after the pores completely disappear, but continues to deviate from the normal grain growth behavior in the case where pore shrinkage rate is highest. To understand this behavior we examined the microstructures of the simulation with the highest pore shrinkage rate and compared them to the others. We found that the initial microstructure for the simulation with the fastest pore shrinkage rate had pores evenly distributes at triple junctions as seen in Fig. 1a. As pore shrinkage was started entire pores would shrink away quickly unpinning a few grain boundaries while most were still pinned. These unpinned grains grew quickly giving abnormal grain growth of a few grains as shown in Fig. 5. At lower pore shrinkage rates, the grain pinning effect of pores is more persistent as well as more even, thus abnormal grain growth did not occur. Furthermore, at the low pore shrinkage rates, the grain size distribution was narrower than that of normal grain growth because of pore pinning.

The changes of average grain sides also indicate that pore shrinkage alters the grain growth, as shown in Fig. 6. For all system, the average grain sides is about 6, which is a consequence of the Euler-Poincare relationship applying to a two-dimensional cellular structures if grains meet only at tri-junctions. The normal grain growth without pore's appearance give the lowest average grain side value and the grain growth in a system which pore only migrates give a value a little higher. The shrinkage of pores effect greatly in the topologies of the microstructure until the pores shrink away completely.

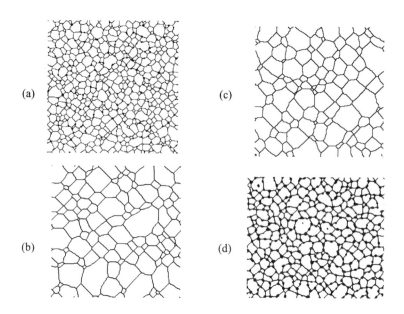

(a)

(c)

(b)

(d)

Fig.5 Microstructures from the fast pore shrinkage simulation at (a) 20,000 and (b) 100,000 MCS showing the development of a broad grain size distribution as a result of uneven pinning by pores. (c) Microstructure of a normal grain growth simulation and (d) grain growth and pore migration simulation with no pore shrinkage to show that both result in a narrower grain size distribution.

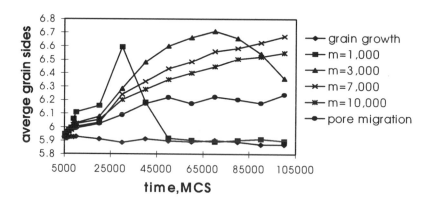

Fig.6 Average grain sides vs. simulation time for different pore migration to pore shrinkage ratios. Pore disappear at 40,000 MCS for m = 1,000.

SUMMARY

The microstructural evolution during the final stage sintering was studied using the Potts model. Grain growth by grain boundary migration, pore migration by surface diffusion and pore shrinkage by grain boundary diffusion were incorporated in the Potts model. An initial porosity of 10% was chosen as a representative microstructure during the final stage sintering. The microstructure evolution shows that pore will migrate along the grain boundaries and coalesce to form larger pores at the triple junctions. Simultaneously, pore shrinkage competed with the pore growth by coalesce especially at the higher shrinkage rates. Simulation results show that the shrinkage of pores and the shrinkage rate effects the topological and grain size distribution. Fast shrinkage results in a broad grain size distribution because some grains are pinned by the pores and will not grow while other grains grow bigger, then the small grains would shrink after all the pores are vanished. In contrast, at the slow pore shrinkage rates, the grain size distribution is narrower because pores hinder grain growth evenly.

Acknowledgment: We thank Thomas Otahal for his assistance with system administration. This work was supported by the US DOE under contract DE-AC-94AL85000. Sandia is a multiprogram laboratory operated by Sandia Corporation, a Lockheed Martin Company, for the USDOE

REFERENCES

[1] M.N. Rahaman, Ceramic Processing and Sintering, Marce Dekker, Inc, 1995

[2] J.E. Burke, " Role of Grain boundaries in sintering", J. Am. Ceram. Soc. 40. 80-85 (1957)

[3] E.A. Holm & V. Tikare, "Simulation of Grain Growth and Pore Migration in a Thermal Gradient", J. Am. Ceram. Soc., 81[3] 480-484 (1998).

[4] E.A. Holm, James A. Glazier, D.J. Srolovitz, G.S. Grest, "Effects of Lattice Anisotropy and Temperature on Domain Growth in the Two-Dimensional Potts Model," Phys. Rev. A, 43 [6] 2662-2668 (1991).

[5] N. Metropolis & S. Ulam, "The Monte Carlo Method", J. Am. Sta. Ass. V44, 335,(1949)

[6] D. J. Srolovitz, M.P. Anderson, P.S.Sahni & G.S. Grest, "Computer simulation of grain growth-II. Grain size distribution, topology, and local dynamics", Acta Metall. V32, No.5, pp792-802 (1984)

[7] V. Tikare & J.D. Cawley, "Application of the Potts Model to Oswald Ripening", J. Am. Cram. Soc., 81[3] 485-491(1998).

[8] R.L. Cobel, "Sintering Crystalline Solid. I. Intermediate and Final State Diffusion Models", J. AM. Ceram. Soc., 32[5] 787-792(1961).

[9] G.N. Hassold, I.Chen, D.J. Srolovitz, "Computer Simulation of Final Stage Sintering: I, Model, Kinetics and Microstructure", J. AM. Ceram. Soc., 73[10] 2857-64(1990).

[10] E.A. Holm, "Modeling Micro-structural Evolution in Single Phase Composite and Two-Phase polycrystals", Dissertation, University of Michigan, 1992.

[11] M. Hillert, "On the theory of Normal and Abnormal Grain Growth", Acta Metall., 13 227-231(1965).

A STUDY OF SELF-REINFORCEMENT PHENOMENON IN SILICON NITRIDE BY MONTE-CARLO SIMULATION

Y. OKAMOTO, N. HIROSAKI AND H. MATSUBARA
Synergy Ceramics Research Laboratory, Fine Ceramics Research Association
2-4-1, Mutsuno, Atsuta-ku, Nagoya-shi, Aichi-ken 456-8587, Japan

ABSTRACT

A grain growth model based on the results of Monte-Carlo simulations is proposed for silicon nitride. The model was derived from the Potts model; in addition, principal characteristics of silicon nitride such as presence of liquid phase and anisotropy of grain growth were introduced. Employing this model, microstructure development of silicon nitride was investigated.

Under certain simulation conditions, several grains grew in preference to other grains, and consequently, a self-reinforced microstructure was produced similar to that of actual silicon nitride. In particular, liquid phase fraction was found to be dominant factor affecting microstructure development.

INTRODUCTION

During sintering of silicon nitride, several grains grow preferentially to form coarse rod-like grains, and these elongated grains act like a toughening second phase. The diameter of elongated coarse grains (typically 5 to 20μm) is extremely large compared to that of the matrix grains ($\cong 1\mu m$). This phenomenon improves mechanical properties of silicon nitride and is often referred to as "self-reinforcement" or "*in-situ* toughening"[1,2]. Although the phenomenon itself is well-known, the mechanisms are not well understood yet. The presence of liquid phase and anisotropy in interface energy inhibit analysis by conventional analytical approaches.

Computer simulation is an alternative method to investigate grain growth behavior. The Potts model[3,4] is one of the most popular grain growth models for Monte-Carlo simulation. The model has been employed, for example, to study abnormal grain growth[5,6], anisotropic grain growth[7], re-crystallization[8,9] and pore elimination[10,11]. Furthermore, Matsubara and Brook have introduced liquid phase into the Potts model[12]. In this new model, the diffusion process in the liquid phase is represented as a random walk. In a simulation of aluminum nitride, the results obtained with this model agreed well with actual materials[13].

In this paper, microstructure development of silicon nitride was studied employing Matsubara's model. In particular, the influence of liquid phase and anisotropic growth was investigated.

SIMULATION METHODOLOGY

Simulation model and algorithm

The simulation model is based on the Potts model[3,4]. Additional features are presence of liquid phase and anisotropy of interface energy, as shown in fig. 1. In this simulation, an important parameter is the ratio of interface energies E_{sl} (solid to liquid) to E_{ss} (solid to solid). Details of the simulation algorithm are given elsewhere[12]. The algorithm essentially consists of four steps: (1) select a cell at random from the simulation lattice (2) change the orientation number of the selected cell. If the cell is adjacent to liquid phase, also change the cell's position via a diffusion process (random walk) until it hits another solid cell. (3) calculate the change of total interface energy (4) if the total energy decreases or does not change, the orientation / position change of the cell is successful.

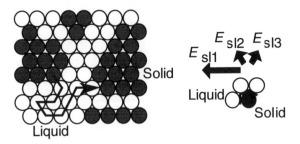

Fig. 1 Simulation model used in this study. Diffusion through the liquid phase is modeled by a random-walk process through liquid phase cells. Anisotropy is introduced by assigning three different interface energies E_{s11}, E_{s12}, E_{s13} for the three possible directions.

Actual silicon nitride contains two types of interface; one is a triple junction and the other is a liquid thin film. In this simulation, thin liquid film was assumed to be present at interfaces between solid cells. Thus the mass transport through a solid to solid interface was assumed to be a solution - reprecipitation process, similar to mass transport through triple junctions. When a solid cell surrounded by solid cells is selected, the new orientation number is assigned from one of the surrounding cells. (In ref.[12], mass transport through a solid to solid interface was performed in the same way as the original Potts model)

Calculation

Simulations were carried out using a 2-dimensional triangular lattice. Periodic boundary conditions were imposed. Orientation numbers Q, ranging from 1 to 64, were distributed to each cell at random. These cells corresponded to solid phase. Furthermore, cells of orientation number 0 were also introduced into the lattice. These cells corresponded to liquid phase.

Interface energies were assigned as follows, E_{ss} =1 (unity), E_{s11}, E_{s12}, E_{s13} ={0.1, 0.1, 0.5}. Since the orientation number of liquid phase was 0 for all cells, no interface energy was attributed to interfaces between liquid phase cells. The direction of higher energy for solid - liquid interface 0.5 was assigned according to the orientation number Q. The size of the simulation lattice was 500 cells by 500 cells. It should be noted that no special grains, e.g. lower interface energy or higher mobility of boundary, were assumed in this simulation.

Simulations with various liquid phase fractions were performed. Each simulation run was carried out for 2,000 Monte-Carlo steps (MCS). Microstructures were recorded every 100 MCS.

RESULTS AND DISCUSSION

Results of simulation at various liquid phase fractions are shown in fig. 2. For small amounts of liquid phase (2.5%, 5%), the microstructure was fairly uniform and no or few coarse grains were observed. In contrast, as the liquid phase fraction increased, several elongated coarse grains appeared in the simulation lattice. The microstructure consisted of large elongated grains and small equiaxed matrix grains. This microstructure was similar to the typical self-reinforced microstructure of actual silicon nitride. The self-reinforcement was found to be most remarkable for 10% liquid phase fraction. When the amount of liquid phase was increased further ($\geqq$15%), microstructures became uniform again, and no preferentially grown grains were observed. This is consistent with experimental reports that the microstructure becomes more uniform for excess amount of liquid phase[14].

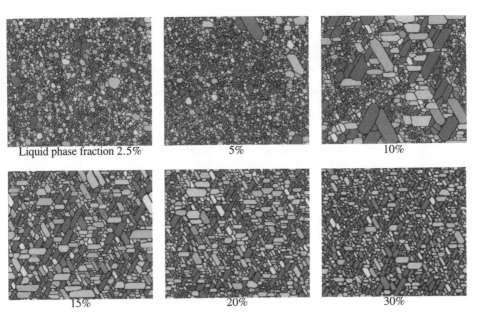

Liquid phase fraction 2.5%   5%   10%

15%   20%   30%

Fig. 2 Microstructures after 2000 MCS. Self-reinforced microstructure was observed only within a particular range of liquid phase fraction.

An important result of the present study is that the self-reinforced microstructure was generated without introducing special grains with preferential grain growth. This result suggests that the liquid phase fraction is one of the most significant factors controlling microstructure, in particular self-reinforcement. Although further inspection and comparison with actual materials are necessary, we believe this simulation model is an appropriate model for studying grain growth of silicon nitride.

CONCLUSION

A new grain growth model for silicon nitride has been proposed based on the Monte-Carlo simulation technique. Characteristic features of silicon nitride, such as presence of liquid phase and grain growth anisotropy was introduced. Grain growth behavior of silicon nitride was studied using the model.

Self-reinforced microstructures were successfully generated. Microstructures varied drastically according to liquid phase amount. Self-reinforcement was observed only within particular range of liquid phase fraction. The amount of liquid phase was found to be a key factor for microstructure control of silicon nitride.

ACKNOWLEDGMENT

This work has been entrusted by NEDO as part of the Synergy Ceramics Project under the Industrial Science and Technology Frontier (ISTF) Program promoted by AIST, MITI, Japan.

REFERENCES

[1] D. E. Wittmer, D. Doshi and T. E. Paulson, in Proceedings of 4th International Symposium on Ceramic Material and Components for Engines, Elsevier Science Publishers, (1992) pp. 594-602.
[2] A. J. Pyzik and D. R. Beaman, *J. Am. Ceram. Soc.*, **7**6[11], 2737-44 (1994)
[3] M. P. Anderson, D. J. Srolovitz, G. S. Grest and P. S. Sahni, *Acta metall.*, **32**[5], 783-91 (1984).
[4] D. J. Srolovitz, M. P. Anderson, P. S. Sahni and G. S. Grest, *Acta metall.*, **32**[5], 793-802 (1984).
[5] G. S. Grest, M. P. Anderson, D. J. Srolovitz and A. D. Rollett, *Scripta Metall. et Mater.*, **24**[4], 661-65 (1990).
[6] A. D. Rollett, D. J. Srolovitz and M. P. Anderson, *Acta Metall. Mater.*, **3**7[4], 1227-40 (1989).
[7] U. Kunaver and D. Kolar, *Acta Metall. Mater.*, **4**1[8], 2255-63 (1993).
[8] D. J. Srolovitz, G. S. Grest and M. P. Anderson, *Acta metall.*, **3**4[9], 1833-45 (1986).
[9] D. J. Srolovitz, G. S. Grest, M. P. Anderson and A. D. Rollet, *Acta metall.*, **36**[8], 2115-28 (1988).
[10] G. N. Hassold, I -W. Chen and D. J. Srolovitz, *J. Am. Ceram. Soc.*, **7**3[10], 2857-64 (1990).
[11] I -W. Chen, G. N. Hassold and D. J. Srolovitz, *J. Am. Ceram. Soc.*, **7**3[10], 2865-72 (1990).
[12] H. Matsubara and R. J. Brook, in "Mass and Charge Transport in Ceramics", Ed. by K. Koumoto, L. M. Shepard and H. Matsubara, Ceramic Transactions vol. 71, The American Ceramic Society, (1996) pp. 403-17.
[13] M. Tajika, H. Matsubara and W. Rafaniello, *J. Ceram. Soc. Japan*, **1**05[11], 928-33 (1997).
[14] C. -J. Hwang and T. -Y. Tien, Materials Science Forum vol. 47, Trans Tech Publications, (1989) pp. 84-109.

THE MICROSTRUCTURE OF PORTLAND CEMENT-BASED MATERIALS: COMPUTER SIMULATION AND PERCOLATION THEORY

Edward J. Garboczi and Dale P. Bentz
National Institute of Standards and Technology
226/B350
Gaithersburg, MD 20899

ABSTRACT

Portland cement-based materials are usually composites, where the matrix consists of portland cement paste. Cement paste is a material formed from the hydration reaction of portland cement, usually a calcium silicate material, with water. The microstructure of cement paste changes drastically over a time period of about one week, with slower changes occurring over subsequent weeks to months. The effect of this hydration process on the changing microstructure can be represented using computer simulation techniques applied to three dimensional digital image-based models. Percolation theory can be used to understand the evolving microstructure in terms of the three percolation thresholds that are of importance in the cement paste microstructure: the set point, capillary porosity percolation, and the percolation of the C-S-H phase.

INTRODUCTION

Percolation theory plays an important role in interpreting and understanding the microstructure of cement-based materials in general. Much can be learned about 3-D systems from 2-D images, about quantities like volume fraction and surface area. However, nothing can be learned about percolation aspects of the microstructure, as percolation is quite different in 2-D and 3-D. Therefore to learn about the 3-D percolation aspects of the microstructure requires some kind of 3-D analysis. This paper shows how a cellular automaton model of cement hydration can give accurate 3-D microstructures and accurate predictions of percolation phenomena.

Cement paste, formed from the reaction of portland cement in water, has more percolation thresholds of importance in it than any other material of which we are aware. There are three percolation thresholds of significance in this material. These all play an important role in relating the microstructure to properties. To understand the significance of these thresholds, one must first know something about percolation concepts, and then something about cement paste microstructure.

PERCOLATION THEORY

The ideas of percolation theory were put together formally by Hammersley in the 1950's [1]. The main idea of percolation theory is to formalize and quantify the meaning of "connectedness" in a random process. Thus, percolation theory can be seen as a branch of topology. Although there are many kinds of percolation, for the point of this paper, we will restrict ourselves to considering the connectedness of random phases in a random multi-phase material. This aspect of percolation theory is called "continuum percolation."

For a material phase, the idea of connectedness is whether a mathematical "ant," moving in only this

Mat. Res. Soc. Symp. Proc. Vol. 529 © 1998 Materials Research Society

phase, can get from one side of the material to the other, in any direction. In a random material, each direction is equivalent. Add to this a process, such that the amount and geometry of a phase is changing in time, and we then have the classic percolation problem: at what volume fraction of the phase will the phase percolate, or form a connected phase? This assumes that the phase starts out as isolated bits, and we are trying to find out when it forms into a connected cluster. We could have the alternative problem: a phase starts out connected, and as material is lost, it gradually becomes disconnected. In either event, the point at which the phase connectivity changes is called the percolation threshold, and the amount of phase present at this point is usually given as the value of the percolation threshold.

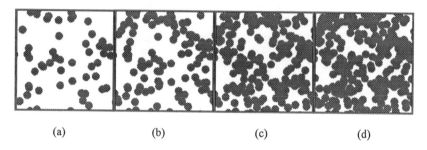

      (a)   (b)   (c)   (d)

Figure 1: Showing four cases of monosize circles (gray) randomly placed on a plane (white), with different area fractions: (a) gray = 0.22, white = 0.78, (b) gray = 0.39, white = 0.61, (c) gray = 0.60, white = 0.40, (d) gray = 0.70, white = 0.30.

A specific 2-D example of these concepts is illustrated in Fig. 1. Here we have a two-phase composite, gray and white, with gray circles being added randomly over time. The center of the gray circles can be located anywhere in the field. Each point of the white field is equally probable, and each gray circle is the same size as all the rest. Figure 1a shows the case where 22% of the material is gray and 78% white. Note that there are already some overlaps between gray circles. The gray phase is discontinuous, and the white phase is continuous. Figures 1b and 1c show the gray phase increasing to 39% and 60%, respectively, but still discontinuous. However, Fig. 1c shows that the gray phase is "almost" continuous, as it is mainly made up of large clusters of overlapping gray circles. Finally, in Fig. 1d, the gray phase has become connected. For this process, careful computer simulation studies have shown the percolation threshold to be about 68% for the gray phase [2]. Note that the percolation threshold for the white phase, the point at which it loses connectivity, is then 32%. In 2-D, for random phases, only one phase can be continuous at a time, so if one phase becomes continuous, the other phase(s) must become or remain discontinuous (non-random phases, like in regular laminar or fibrous composites, can have many phases continuous at the same time).

This is not the case in 3-D, however. This is the main reason why the study of 2-D slices cannot tell us much, if anything, about 3-D percolation. Figure 2a shows a slice through the three dimensional equivalent of Fig. 1, randomly placed monosize spheres. The volume fraction of the sphere phase in the figure is 40%, which is beyond the percolation threshold for the sphere phase [3]. Therefore, in 3-D, the spheres are percolated, while in this 2-D slice, they clearly are not. In 3-D, an infinite number of phases can be percolated at the same time, and so when one phase becomes connected, the other phases can retain their connectivity [4].

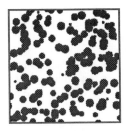 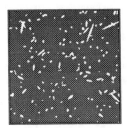

Figure 2: Showing: (a) a slice through a 3-D system of 40% by volume randomly placed monosize spheres, and (b) a slice through a 3-D system of 7% by volume randomly placed prolate (10:1 aspect ratio) ellipsoids of revolution. In 3-D, both sets of objects are percolated, while in the 2-D slices, they are obviously not percolated.

Figure 2b further illustrates this point with a system of 3-D randomly placed ellipsoids (shown in white). The ellipsoids are prolate ellipsoids of revolution, with an aspect ratio (long to short semi-axis length) of 10. There are about 7% by volume of the ellipsoids. In the 2-D slice shown in Fig. 2b, the ellipsoids look very much disconnected, but a numerical study [3] showed that most of these do form a connected phase in 3-D, with less than half still being isolated.

The ellipsoid example shows us that at percolation, only part of a phase can initially make up the connected cluster, with the other parts connected over time as more of the phase is added. We define the "fraction connected" by counting how much of the phase is contained in a percolated cluster, and dividing by the total amount of that phase present. If the entire phase is isolated, then this ratio is zero. If all of the phase makes up a percolated cluster, then this ratio is 1.

The concept of "fraction connected" is illustrated by Fig. 1c, reproduced below as Fig. 3. In Fig. 3, the section of the white phase enclosed by the black line is isolated, even though most of the white phase is still percolated. This would cause the fraction connected of the white phase to be less than one, but still greater than zero.

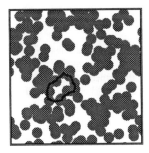

Figure 3: Area fractions: gray = 0.60, white = 0.30. Black line encloses a region of white phase that isolated from the rest of the continuous white phase.

This concept is more clearly illustrated by Fig. 4, which shows two systems of randomly placed ellipses, 550 in Fig. 4a, and 700 in Fig. 4b [5]. The ellipses were randomly oriented in either the x or the y directions. The right-hand image in each set shows the ellipses that are accessible from the top in white. In Fig. 4a, clearly 550 ellipses, or an area fraction of about 40%, were not enough for percolation of the ellipses to occur. In Fig. 4b, 700 ellipses or an area fraction of about 47% were placed, and the companion image shows that there is now a percolated pathway from top to bottom. A substantial number of ellipses were not contained in this pathway, however, giving a fraction connected of about 0.60.

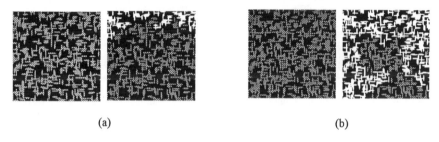

(a)   (b)

Figure 4: Showing ellipses (gray) randomly placed on a black background. The right-hand image in each pair shows how much of the ellipse phase is accessible (white) from the top down: (a) 40% ellipses, (b) 47.5% ellipses. In (b), the fraction connected of the ellipse phase is about 0.60.

MICROSTRUCTURE OF PORTLAND CEMENT PASTE

The starting point for portland cement paste (cement + water), cement powder, is obtained by grinding cement clinker with about 5% of gypsum (calcium sulphate dihydrate). The gypsum is added to moderate the hydration process. The cement clinker is manufactured by firing carefully proportioned and interground mixtures of limestone, clay, and iron ore, or other sources of CaO, SiO_2, Al_2O_3, and Fe_2O_3. After grinding, the cement powder consists of multi-size, multi-phase, irregularly shaped particles generally ranging in size from less than 1 μm to about 100 μm, with an average diameter of about 15-20 μm .

The scanning electron micrograph of cement particles in Fig. 5 shows that even before hydration takes place, the cement powder is itself a random composite, with even most particles themselves being multi-phase composites. When the cement is mixed with water, hydration reactions occur which ultimately convert the water-cement suspension into a rigid porous material, which serves as the matrix phase for mortar and concrete. The degree of hydration at any time is the fraction of the cement that has reacted with water, and is often denoted by the symbol α. The ratio of water to cement in a given mixture is specified by the water to cement ratio (w/c), which is the ratio of the mass of water used to the mass of cement used. The various chemical and mineral phases within the cement powder hydrate at different rates, depending on their size and composition, and interact with one another to form various reaction products. Some products deposit on the remaining cement particle surfaces (surface products) while others form as crystals in the water-filled pore space between cement particles (pore products). For simplicity, and because it still correctly captures the main features of the pore structure, cement paste can be thought of as consisting of four phases: 1) unreacted cement grains, 2) surface products like C-S-H (calcium- silicate-hydrate), which is the main amorphous "glue" of cement paste and is itself nanoporous, 3) pore products like CH (calcium hydroxide), which forms crystalline masses, and 4) capillary pore space, which is the remaining water-filled space around the cement grains and hydration products. Surface products grow outward from the unreacted cement particles, while pore products nucleate and grow in the pore space.

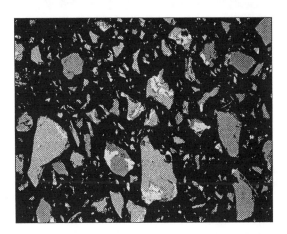

Figure 5: Back-scattered scanning electron micrograph of a portland cement powder embedded in epoxy. The gray scales show the random mineral composition of the grains. The main phase is calcium tri-silicate = C_3S.

While images of both initial and hydrated cement microstructures can be experimentally obtained in two dimensions, acquiring quantitative three-dimensional information is much more difficult. It is for this reason that computer models of the 3-D microstructure development of cement paste have been developed. The actual process of cement hydration, for the purposes of modelling the development of microstructure, can be broken down into three parts: 1) dissolution from the original cement particle surfaces, 2) diffusion within the available pore space, and 3) reaction with water and other dissolved or solid species to form hydration products. Each of these processes may be conveniently simulated using cellular automaton-type rules as has been previously described [6,7]. Figure 6 shows four steps of simulated hydration for a C_3S cement paste in 2-D. The original particle shapes are taken from a backscattered SEM image.

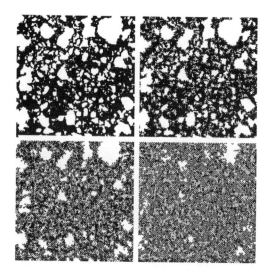

Figure 6: Four stages of hydration in a microstructural model of C_3S hydration. The degrees of hydration are: top left--0, top right--20 %, bottom left--50%, bottom right--87%. White = unreacted cement, light gray = CH, dark gray = C-S-H, and black = porosity.

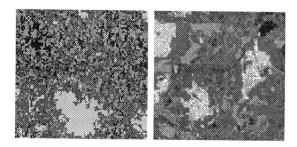

Figure 7: Qualitative comparison between: (right) backscattered scanning electron micrograph of real cement paste and (left) model version with realistic gray scale.

A qualitative comparison between a real backscattered scanning electron micrograph of a cement paste and a model version, in which a gray scale has been used to match that of the real picture, is shown in Fig. 7. Reasonable comparison with experiment is obtained by the model.

This brief description of the chemical hydration process, which is the basis of the microstructural formation of cement paste, of course glosses over a number of chemical details, many of which are not clearly understood. However, this simple description is sufficient to be able to go on and investigate the various important percolation thresholds that occur in cement paste.

PERCOLATION THRESHOLDS IN CEMENT PASTE

Set point

Immediately after mixing together water and cement, a dense suspension of cement particles in water is achieved. Set is the phenomenon where the dense suspension, which is a viscoelastic liquid with a high plastic viscosity and a finite but small yield stress [8], turns into a viscoelastic solid, with a finite long-time shear modulus. This is a variation of what is well-known in the polymer literature as a sol-gel transition.

Set is a percolation phenomenon, since it is achieved when the solid cement grains become linked together with enough hydration product, mainly C-S-H, so that they form a rigid backbone with a finite long-time or zero frequency shear modulus. This process is illustrated schematically in Fig. 8. On the left, cement grains (light gray) form an isolated phase when they are first mixed into water. As time goes by, however, and hydration products like C-S-H (dark gray) form, the cement grains gradually become linked together by C-S-H to form a continuous phase [8,9]. The phase that actually percolates is a composite phase of cement grains plus C-S-H. Set usually occurs at a low degree of hydration, about $\alpha = 0.02$-0.08 [8]. This has been successfully predicted by the cement paste microstructure model [9].

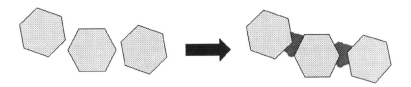

Figure 8: Schematic illustration of the setting process for cement paste. The light gray grains are cement, the dark material linking the cement grains is C-S-H, and the dark arrow shows the elapse of time, usually about 3-6 hours at ordinary temperatures using typical cements.

Capillary porosity percolation and transport phenomena

As hydration products consume both water and cement and fill in the capillary pore space, the capillary porosity is reduced. Just as in the 2-D gray-white case illustrated in Fig. 1 above, the connectivity of the capillary pore space (white) is gradually reduced as it is filled-in by hydration products (gray). This process can be followed experimentally by measuring transport properties like electrical conductivity [10,11], and theoretically using the cement paste microstructure model [12]. Figure 9 shows the model results for the connectivity of the capillary pore space for a variety of w/c ratios, for pure C_3S cement pastes. Figure 9a shows how this percolation is a function of degree of hydration as expected, with w/c ratio pastes above about 0.6 having their capillary pore spaces remaining percolated throughout hydration. Figure 9b reports this data vs. capillary porosity, and shows that all the data roughly fall on a single curve, with a percolation threshold of about 18%

porosity. The pastes with w/c ratio of 0.6 or more still follow this curve, but since their porosity never can reach the percolation threshold (too much water initially for the amount of cement used), their capillary pore space remains percolated. The threshold has been found to have a small dependence on cement particle size distribution and cement chemical type, and ranges from 0.18-0.22 volume fraction.

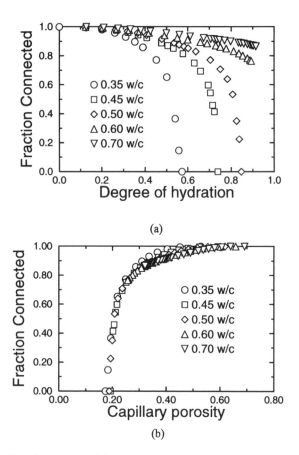

(a)

(b)

Figure 9: Fraction connected for capillary porosity for various w/c ratio cement pastes: (a) plotted vs. degree of hydration, and (b) plotted vs. capillary porosity.

C-S-H percolation: Cement paste at –40°C

The cement paste microstructure model discussed above predicts that the C-S-H phase in a cement paste will itself percolate at a volume fraction of 15-20% [6]. Figure 10 shows a schematic view of this process.

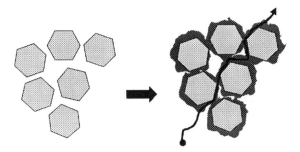

Figure 10: Schematic illustration of the C-S-H percolation process (light gray = cement, dark gray = C-S-H, arrow indicates the elapse of time, about 12-24 hours, and the curved arrow shows continuity of the C-S-H phase).

Experimentally, the only way one can investigate this phase is through transport studies using electrical conductivity, for example. However, the main transport paths through cement paste are through the much larger capillary pores. These only become discontinuous at later stages of hydration, as was seen above, so that transport at that time, even though it is mainly through C-S-H pores, does not tell us anything about C-S-H at earlier stages of hydration, when its volume fraction is near its predicted percolation threshold.

Upon dropping the temperature of the cement paste to –40°C, it has been found that the larger capillary pores freeze, thus sharply reducing the conductivity of the ionic pore solution contained in them, while the solution in the smaller nano-scale pores in the C-S-H phase does not freeze. At this temperature, it is then found that the C-S-H phase becomes the main transport pathway [13]. In an electrical conductivity experiment, there is only conduction through the sample when the C-S-H phase is continuous. Figure 11 shows the results of such measurements, for two different w/c ratios and for model predictions and experiments [13]. When plotted against the volume fraction of C-S-H present in the material, all the data falls on a single curve and shows a percolation threshold of about 15-20% volume fraction of C-S-H, as predicted by the model [6,13].

DISCUSSION AND CONCLUSIONS

Cement paste, a micrometer scale material, has the most percolation thresholds of significance of any material that we know. These are three in number, and include the set point, the C-S-H percolation threshold, and the capillary porosity percolation threshold.

If one goes down in length scale, to the nanometer length scale, one sees the structure of C-S-H, which is itself a nanoporous material. There does not seem to be any percolation phenomena at this length scale. These nano-scale pores seem to be always connected and allow transport.

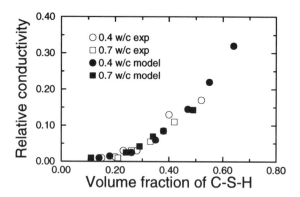

Figure 11: Showing the conductivity of the cement paste (at 40°C) relative to the conductivity of C-S-H, plotted vs. the volume fraction of C-S-H.

If we go up in length scale, to the millimeter length scale, cement paste is used as the matrix for concrete, surrounding the rock and sand inclusions. There is a percolation phenomenon of importance at this length scale. Because of the particulate nature of cement paste, at first mixing the cement grains do not pack as well around the inclusions as out in the bulk paste regions. This gives an interfacial transition zone (ITZ) around each inclusion that has less cement, and, therefore, more porosity than the bulk paste. This is often called the wall effect [14]. In concrete, we can then think of each inclusion being surrounded by a 10-30 μm (the typical size of a cement particle) thick shell, representing the ITZ, which has a higher porosity and larger pores than does the rest of the cement

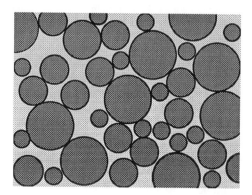

Figure 12: Schematic picture of many sand and rock grains (dark gray) packed into a cement paste matrix (light gray), each surrounded by interfacial transition zones (black).

paste. Figure 12 shows a 2-D schematic picture of how these regions look in a concrete. In Fig. 12, the ITZ regions do not percolate. However, remember that a 2-D picture cannot give 3-D percolation information. These ITZ regions have higher transport properties like ionic diffusivity and fluid permeability than does the bulk paste, so it is important to know whether these rims or shells form a percolating phase in 3-D. This geometry is a classic percolation problem--the hard-core soft-shell problem [15]. For most concretes, the ITZ regions usually do percolate, depending on the volume fraction of rocks and sand present and the thickness of the ITZ regions [16].

So we see that percolation ideas, along with composite ideas, explain and link the microstructure of cement-based materials across many length scales, from nanometers to milllimeters and beyond, and are particularly important at the micrometer length scale of cement paste and the millimeter length scale of concrete. More detailed exploration of these multi-scale ideas, combined with percolation and composite theory ideas, can be found at http://ciks.cbt.nist.gov/garboczi/, "An electronic monograph: Modelling the structure and properties of cement-based materials."

ACKNOWLEDGEMENTS

I would like to thank the NSF Center for Advanced Cement-Based Materials (ACBM), and the Partnership for High Performance Concrete project at the National Institute of Standards and Technology, for supporting this work

REFERENCES

[1] J.M. Hammersley, Proc. Cambridge Phil. Soc. 53, 642 (1957).

[2] E.J. Garboczi, M.F. Thorpe, M. DeVries, and A.R. Day, Phys. Rev. A 43, 6473-6482 (1991).

[3] E.J. Garboczi, K.A. Snyder, J.F. Douglas, and M.F. Thorpe, Phys. Rev. E 52, 819-828 (1995).

[4] D. Stauffer and A. Aharony, Introduction to Percolation Theory (2nd ed.) (Taylor and Francis, London, 1992).

[5] E.J. Garboczi, D.P. Bentz, and N.S. Martys, "Digital imaging and pore morphology," in Experimental Methods for Porous Materials, edited by P. Wong (Academic Press. New York, 1999).

[6] D.P. Bentz and E.J. Garboczi, Cement and Concrete Research 21, 325-344 (1991).

[7] D.P. Bentz, J. Amer. Ceram. Soc. 80, 3-21 (1997).

[8] Y. Chen and I. Odler, On the Origin of Portland Cement Setting, Cem. Conc. Res. 22, 1130-1140 (1992).

[9] D.P. Bentz, E.J. Garboczi, and N.S. Martys, in Proceedings of 1994 NATO/RILEM Workshop The Modelling of Microstructure and Its Potential for Studying Transport Properties and Durability, edited by H.M. Jennings, pp. 167-186. See also http://ciks.cbt.nist.gov/garboczi/, Appendix 1.

[10] R.T. Coverdale, B.J. Christensen, T.O. Mason, H.M. Jennings, E.J. Garboczi, and D.P. Bentz, J. Mater. Sci. 30, 712-719 (1995).

[11] B.J. Christensen, T.O. Mason, H.M. Jennings, D.P. Bentz, and E.J. Garboczi, in *Advanced Cementitious Systems: Mechanisms and Properties*, edited by F.P. Glasser, G.J. McCarthy, J.F. Young, T.O. Mason, and P.L. Pratt (Materials Research Society Symposium Proceedings Vol. 245, Pittsburgh, 1992), pp. 259-264.

[12] E.J. Garboczi and D.P. Bentz, J. Mater. Sci. **27**, 2083-2092 (1992).

[13] R.A. Olson, B.J. Christensen, R.T. Coverdale, S.J. Ford, G.M. Moss, H.M. Jennings, T.O. Mason, and E.J. Garboczi, J. Mater. Sci. **30**, 5078-5086 (1995).

[14] K.L. Scrivener and E.M. Gartner, in *Bonding in Cementitious Composites*, ed. S. Mindess and S.P. Shah (Materials Research Society, Pittsburgh, 1988), pp. 77-85.

[15] I. Balberg, Philo. Mag. B **56**, 991 (1987); I. Balberg and N. Binenbaum, Phys. Rev. A **35**, 5174 (1987); S.B. Lee and S. Torquato, J. Chem. Phys. **89**, 3258 (1988); S. Torquato, J. Chem. Phys. **85**, 6248 (1986).

[16] D.N. Winslow, M.D. Cohen, D.P. Bentz, K.A. Snyder, and E.J. Garboczi, Cement and Concrete Research **24**, 25-37 (1994).

CELLULAR AUTOMATON MODELING OF ALLOY SOLIDIFICATION USING LOCAL ANISOTROPY RULES

R. E. NAPOLITANO* and T. H. SANDERS, JR.**
*Metallurgy Division, National Institute of Standards and Technology, Gaithersburg, MD, 20899
**Materials Science and Engineering, Georgia Institute of Technology, Atlanta, GA, 30332

ABSTRACT

The evolution of dendritic morphology is simulated for a binary alloy using a two-dimensional cellular automaton growth algorithm. Solute diffusion is modeled with an alternate-direction implicit finite difference technique. Interface curvature and kinetic anisotropy are implemented through configurational terms which are incorporated into the growth potential used by the automaton. The weighting of the anisotropy term is explored and shown to be essential for overcoming grid-induced anisotropy, permitting more realistic development of dendritic morphologies. Dendritic structures are generated for both uniform and directional cooling conditions.

INTRODUCTION

Virtually every microstructural feature in a cast alloy component is related to the interface morphology present during solidification. Conventional solidification theory, describing dendrite tip kinetics and mushy zone phenomena, provides a means for estimating overall microstructural parameters for an alloy solidified under steady-state conditions. Microstructural prediction for actual components, which are finite in size and complex in geometry, requires a description of the evolution of the interface morphology under transient conditions and geometrical constraint and an understanding of how the structure propagates through the mold. Any useful simulation tool must be equipped to handle these circumstances within the limits of computational feasibility.

Solidification involves the growth of a solid phase into a liquid phase, where the growth is controlled by the continuity of thermal and solutal fields along with energetic conditions at the interface. A major problem encountered in solidification modeling, however, is that the controlling phenomena operate at different size scales. Depending upon the goal of a particular model, the mechanisms which must be addressed in the modeling of a solidification process may vary considerably. In general, the issues involved can be divided into two classes. At the macroscopic level, the transport of heat and solute govern the conditions which are present at the advancing interface. These are handled with the appropriate differential equations which can be discretized and solved using various techniques. At the microscopic level, the response of the interface to the instantaneous conditions must be described. This includes the motion of the interface and the associated redistribution of heat and solute. In recent years, various phase-field[1-6] and cellular automaton (CA)[7-14] models have been used to simulate many features of dendritic solidification.

Phase-field methods involve describing the interface with one or more order parameters which may vary continuously between two constant values, indicating the respective phases. The energy or entropy of the system is described in Cahn-Hilliard fashion using a double-well potential with minima associated with the solid and liquid values of the order parameters. The evolution of the continuous phase-field is governed by the dissipation of energy or by the production of entropy. These methods have been used to successfully model many aspects of dendritic solidification at the microscopic level.[1-6] The technique is computationally limited, however, by the requirement of an interface which is both very thin and well resolved. In some cases, this limitation does not permit the implementation of realistic conditions.

Cellular automata are discrete lattice, rule-based growth algorithms developed to study the evolution of self-organizing systems of identical components. The utility of these methods is only beginning to be appreciated, and the use of automata for solidification modeling has been limited. The CA technique was first applied to the phenomenon of dendritic crystal growth in 1985 by Packard.[7] This model produced equiaxed dendritic shapes in two dimensions for a pure substance by considering heat flow, interface curvature, and latent heat generation. Brown and Spittle[8] simulated the formation of dendritic grain structures in a pure metal by assuming an initial distribution of nuclei and using a Monte-Carlo algorithm for further nucleation and growth. Growth was based on minimization of energy, which included a bulk term for each phase and an interfacial term. By varying the boundary conditions, the qualitative effects of melt superheat and mold temperature on the columnar to equiaxed transition, were reproduced. In a subsequent model, which was a direct extension of the original Packard model, the evolution of dendrite morphology in an initially uniformly undercooled pure metal was simulated.[9] Rappaz and Gandin have modeled the formation of grain structures during solidification by using a CA growth algorithm based on a prescribed growth relationship governing the dendrite tip kinetics, where the velocity is an explicit function of the temperature. Thus, by assuming that all growth is fully dendritic, the growth of the dendrite envelope is modeled by tracking only the tips. The CA for growth is coupled to a finite element (FE) thermal calculation which allows for the liberation of heat based on a solid fraction-temperature function, truncated at the temperature of the dendrite tip. The temperature at each tip is obtained from the FE calculation and used by the CA algorithm to update the shape of the solid. By incorporating various aspects of nucleation and fluid flow, several features of solidification structures have been modeled quantitatively.[10-15]

In previous work by the authors, a cellular automaton approach has been used to mimic morphological instability and to simulate dendritic solidification structures for a range of directional growth conditions.[16,17] Additionally, it has been shown that anisotropic configurational parameters are necessary to produce stable dendritic morphologies.[17] The objective of the current work is to advance this approach by exploring the effects of a configurational term for kinetic anisotropy.

MODEL DESCRIPTION

The simulation domain is a square mesh where each cell is assigned a value for temperature, composition, and phase. Temperature and composition are continuous field variables while the phase is a discrete binary variable. The overall operation of the model involves coupling the evolution of the temperature and solute field with the motion of the phase boundary. The solute field is updated using an alternate-direction-implicit finite-difference (ADI-FD) method with periodic lateral boundaries. For the simulations presented here, the thermal field is specified as either directional or uniform. The uniform conditions are implemented by imposing a fixed cooling rate on a field of uniform temperature. Conditions for directional solidification are implemented by employing a uniform temperature gradient and a constant isotherm velocity. In the directional case, the simulation frame is permitted to move in order to follow the front over distances much greater than the domain itself.

The morphology of the solid is evolved using a CA technique which is intended to simulate the morphological evolution of an advancing solidification front by including the relevant physical factors such as temperature, composition, and interface curvature. The contribution from each is incorporated into a growth function which can be applied to every cell within the domain. Each cell is, thus, free to grow as the local conditions allow, and no explicit distinction is made for dendrite tips or other features. The morphology is updated by comparing random numbers to a growth probability $p(\phi)$, where ϕ is a growth potential computed from the local temperature, composition, and morphology. This method naturally imparts a random noise component to the structure, facilitating the fluctuations necessary to initiate morphological changes. The function p, which is essentially a velocity function, must be bounded by zero and one and should vary monotonically with the growth potential, ϕ. The following function is chosen:

$$p = 1 - \exp[-(\frac{\phi}{\kappa})^{\eta\kappa}] \tag{1}$$

where κ and η can be used to control the shape of the function over the potential domain. The defining characteristics of the model, therefore, are the formulation of the potential ϕ and the description of the constants η and κ, in terms of the physical properties of the alloy.

For a site at position (i,j), the growth potential at time k is defined as the kinetic undercooling:

$$\phi_{ij}^{k} = \Delta T - (\Delta T_C + \Delta T_R) \tag{2}$$

where ΔT is the total undercooling, and the subscripts C and R indicate compositional and curvature related contributions, respectively. With knowledge of the phase diagram, ΔT_C at any cell can be calculated based on the local temperature and composition. With a configurational term for ΔT_R, (2) becomes:

$$\phi_{ij}^{k} = T_m - mC_{ij}^{k} - f(S_{ij}^{k}) - T_{ij}^{k} \tag{3}$$

where T_m is the melting temperature, m is the liquidus slope, C_{ij}^{k} is the solute concentration, T_{ij}^{k} is the temperature, and S_{ij}^{k} is the configuration of the neighborhood about cell (i,j) at time k. The configurational contribution to undercooling is computed as:

$$f(S_{ij}^{k}) = \Gamma \frac{(\alpha_r H_r + \alpha_a H_a + \beta)^{\xi}}{\Delta z} \tag{4}$$

where Γ is the Gibbs-Thompson coefficient, β and ξ are modeling constants, and Δz is the grid resolution. The H_i values are contributions due to interfacial curvature (i=r) and anisotropy (i=a), weighted by the coefficients α_r and α_a, respectively. In terms of the first and second nearest neighbors, these can be expressed as follows.

$$H_r = 2 * [\omega(2-\epsilon_1) + (1-\omega)(2-\epsilon_2)] \tag{5}$$

$$H_a = (\epsilon_1 - 2)\delta(\epsilon_1, \epsilon_2) + (\epsilon_1 - \epsilon_2)[\epsilon_2 - 2H'(\epsilon_2 - 2)]$$

where ω is a weight factor between 0 and 1, ϵ_i is the number of solid i^{th} nearest neighbors, δ is the Kronecker delta, and H' is the Heavyside function.

At this point, we have a method for moving the interface based on the local value of ϕ, which can be computed from the field variables and solid morphology. To determine the kinetic resistance imparted by (1), we evaluate $dp/d\phi$ at a characteristic undercooling of κ.

$$\frac{dp}{d\phi}\Big|_{\kappa} = \frac{\eta}{\exp(1)} \equiv \eta' \tag{6}$$

Generally, the velocity of an interface is related to the kinetic undercooling through a proportionality constant:

$$V = M\Delta T \qquad (7)$$

where,

$$M = J_0 v \, \frac{\Delta s_f}{RT} \, \exp(\frac{-Q}{RT})$$

J_0 is vibrational frequency, v is atomic volume, Q is activation energy for diffusion, R is the gas constant, and Δs_f is the entropy associated with solidification.[18] In the model, the velocity is given by the product of the growth probability, the time step frequency, and the cell size of the grid:

$$V = p \, \frac{\Delta z}{\Delta t} = p V_{max} \qquad (8)$$

where V_{max} is the limiting velocity. Combining (6) and (7):

$$p = \frac{M\Delta T}{V_{max}} \qquad (9)$$

Considering (1) and (8), η' can be computed by relating the differentials:

$$\eta' = \frac{dp}{d\phi} = \frac{dp}{d(\Delta T)} = \frac{M}{V_{max}} \qquad (10)$$

We now have a technique for calculating the value of p at each cell, based on the relevant physical parameters. To update the solid morphology, a random number ($0 \leq r \leq 1$) is generated for each cell at each time step for comparison with p. If $p_{ij}^k > r_{ij}^k$, the cell is set to solid. The composition is then set to kC_{ij}^k, and the excess solute $(1-k)C_{ij}^k$ is distributed among the available neighboring cells.

MODELING RESULTS

All results presented in this section were obtained using input parameters associated with Al-4.5 wt% Cu. The effect of the anisotropic contribution was explored by examining the growth of a single seed in the center of a uniform melt which is cooled at a constant rate. The grid resolution for these calculations is 1.5×10^{-5} m. The initial seed is circular in shape, with a diameter of 3.0×10^{-4} m. The effect of H_a on the evolution of growth shape from the seed is shown for three different values of H_a in Figure 1. In Figure 1a, the value of H_a is not sufficient to overcome the anisotropy introduced by the grid, which dictates the initial square growth shape. For this shape, diffusive effects at the corners promote rapid growth and the shape evolves as shown. In Figure 1b, the value of H_a is high enough to result in an initial growth shape with {11} interfaces. Once again, the corners grow rapidly and the shape evolves into a dendritic structure with side branches. In Figure 1c, the value of H_a is, again, high enough to promote {11} interfaces in the initial growth shape. As the shape evolves, however, it becomes clear that H_a is so high that other orientations have been almost completely suppressed, resulting in unnatural structures. A more advanced stage of morphological evolution for the intermediate value of H_a is shown in Figure 2.

In the case of directional growth, the effect of the anisotropy parameter is equally important. Figure 3 shows the initial structural evolution during directional growth, using three different values for H_a. For H_a=0.1, the anisotropy imparted by the grid promotes {01} interfaces. This suppresses the development of curved cell tips. Cell fronts remain relatively flat until the undercooling becomes high enough to overcome the grid effect. This results in an abrupt and unnatural transition in morphology during the evolution of the dendritic front, as shown in Figure 3a. For H_a=0.3, the structure is no longer dominated by {01} interfaces, and it evolves smoothly into a dendritic front, as shown in Figure 3b. When H_a is increased further to a value of 0.5, as in Figure 3c, the {11} interfaces dominate the structure, resulting in pointed dendrite tips. Figure 4 shows the near-steady-state morphology resulting from directional growth using the intermediate value of H_a.

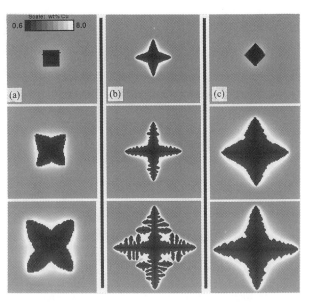

CONCLUSIONS

The anisotropy imparted by the grid has a significant effect on the resulting morphology. The

Figure 1. A comparison between simulated growth morphologies for a circular seed in a uniform melt: (a) Hs=0.1, (b) Hs=0.3, (c) Hs=0.5. Domain = 3 x 3 mm. Mesh = 200 x 200.

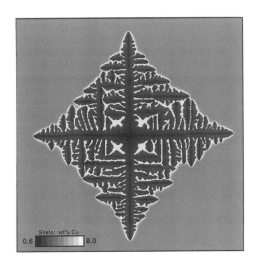

Figure 2. Later stage of growth for the simulation shown in Figure 1b. Domain = 6 x 6 mm. Mesh = 400 x 400.

incorporation of kinetic anisotropy, even through a simple configurational term, is essential for the simulation of realistic dendritic patterns using a cellular automaton approach.

REFERENCES

1. R. Kobayashi; *Physica D*, **63** (1993) 410-423.
2   A.A. Wheeler, B.T. Murray, and R.J. Schaefer; *Physica D*, **66** (1993) 243.
3. S.L. Wang, R.F. Sekerka, A.A. Wheeler, B.T. Murray, S.R. Coriell, R.J. Braun, and G.B. McFadden; *Physica D*, **69** (1993) 189-200.
4. A.A.Wheeler, W.J. Boettinger, and G.B. McFadden; *Physical Review E*, **47** (1993) 1893.
5. J.A. Warren and W.J. Boettinger; *Acta Metall. Mater.*, **43** (1995) 689-703.

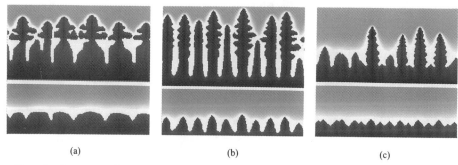

| (a) | (b) | (c) |

Figure 3. A comparison between simulated morphologies developing from a flat seed along the boundary, with directional cooling conditions: (a) Hs=0.1, (b) Hs=0.3, (c) Hs=0.5. Domain = 3 x 3 mm. Mesh = 200 x 200.

6. W.J. Boettinger and J.A. Warren; *Met. Mater. Trans.*, **27A** (1996) 657-669.
7. N.H. Packard; *Proceedings of the First Int. Symposium for Science on Form*, (Y. Katoh, R. Takaki, J. Toriwaki, and S. Ishizaka, Eds.), KTK Scientific Publishers (1986).
8. S.G.R. Brown and J.A. Spittle; *Scripta Metallurgica*, **27** (1992) 1599-1603.
9. S.G.R. Brown, T. Wiliams, and J.A. Spittle; *Acta Metall. Mater.*, 42 (1994) 2893-2898.
10. M. Rappaz and Ch.-A. Gandin; *Acta Metall. Mater.*, **41** (1993) 345-360.
11. Ch.A. Gandin, M. Rappaz, and R. Tintillier; *Met. Trans.*, **24A** (1993) 467-479.
12. Ch.A. Gandin, M. Rappaz, and R. Tintillier; *Met. Trans.*, **25A** (1994) 629-635.
13. M. Rappaz, Ch.A. Gandin, and R. Sasikumar; *Acta Metall. Mater.*, **42** (1994) 2365-2374.
14. Ch.-A. Gandin and M. Rappaz; *Acta Metall. Mater.*, **42** (1994) 2233-2246.
15. Ch.-A. Gandin, Ch. Charbon, and M. Rappaz, *ISIJ International*, **35** (1995) 651-657.
16. R.E. Napolitano and T.H. Sanders, Jr., *Proc. Int. Symp. on Processing of Metal and Advanced Materials: Modeling, Design, and Properties*; B.Q. Li, E d., TMS (1998) 63-74.
17. R.E. Napolitano and T.H. Sanders, Jr., Proc. *Third Pacific Rim International Conference on Advanced Materials and Processing*; TMS (1998).
18. K.A. Jackson, *Journal of Crystal Growth*, 3/4 (1968) 507-517.

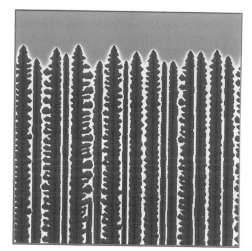

Figure 4. Later stage of growth for the simulation shown in Figure 3b. Domain = 6 x 6 mm. Mesh = 400 x 400

SIMULATION OF THE TEXTURE EVOLUTION OF ALUMINIUM ALLOYS DURING PRIMARY STATIC RECRYSTALLIZATION USING A CELLULAR AUTOMATON APPROACH

V. MARX, G. GOTTSTEIN
Institut für Metallkunde und Metallphysik, RWTH Aachen, Kopernikusstr. 14,
52056 Aachen, Germany, *marx@imm.rwth-aachen.de*

ABSTRACT

A 3D model has been developed to simulate both primary static recrystallization and recovery of cold worked aluminium alloys. The model is based on a modified cellular automaton approach and incorporates the influence of crystallographic texture and microstructure in respect to both mechanisms mentioned above. The model takes into account oriented nucleation using an approach developed by Nes for aluminium alloys. The subsequent growth of the nuclei depends on the local stored energy of the deformed matrix (i.e. the driving pressure) and the misorientation between a growing nucleus and its surrounding matrix (i.e. the grain boundary mobility). This approach allows to model preferred growth of grains that exhibit maximum growth rate orientation relationship, e.g. for aluminium alloys a 40° <111> relationship with the surrounding matrix. The model simulates kinetics, microstructure and texture development during heat treatment, discrete in time and space.

INTRODUCTION

The mechanical properties of metallic materials are strongly affected by their microstructure. To optimise these properties in a processed material, control of the microstructural evolution is of high importance. The most important tools to achieve this control are deformation and heat treatment of a material. Heat treatment of cold worked materials leads to recovery and in particular recrystallization processes. Despite an abundance of reliable data on deformation and annealing processes, primary static recrystallization of heavily deformed metals is still poorly understood. Due to the complexity of the involved processes analytical treatments so far have been unable to properly account for the microstructural evolution and associated texture change.

Correspondingly, several attempts have been made to numerically simulate these processes on a computer. Our approach to these problems uses a 3D cellular automaton model for the simulation of primary static recrystallization and static recovery. Due to the flexibility of a cellular automaton approach, the model is capable of tackling a large number of different nucleation scenarios. Furthermore, the influence of different grain boundary velocities on the evolution of microstructure and texture is incorporated in the model.

THEORY

The model for the description of primary static recrystallization is based on a cellular automaton approach. The model considers the local microstructure and microtexture at length scales ranging from a few subgrains up to several hundred micrometers. The model yields data on the kinetics, the microstructural and textural evolution during annealing of cold worked metals and alloys. A detailed description of the model will be published in [1,2]. In this contribution we shall confine our consideration to texture evolution

Mat. Res. Soc. Symp. Proc. Vol. 529 © 1998 Materials Research Society

The cellular automaton model used consists of a 3D cubic grid onto which the deformed microstructure is mapped. The model is discrete in space and time. The properties, i.e. the crystallographic texture, and the dislocation density of each cell embedded in this grid are stored. During each step of the model three physical processes, namely recovery, nucleation and growth of the nuclei, are sequentially simulated.

To simulate nucleation, spheres of a given radius are arranged within the deformed matrix. The cells inside of these spheres are assigned a crystallographic orientation depending on the assumed nucleation mechanism. The kinetics of nucleation are either incorporated as site saturated nucleation or as a constant nucleation rate. During growth of the nuclei these spheres grow into the surrounding deformation matrix. The velocity of the grain boundaries depends on their mobility and on the driving force. The first property is expressed in terms of the misorientation between the nucleus and the matrix. The second property is expressed in terms of the difference in dislocation density. All cells at the surface of growing grains are then displaced into the surrounding deformation matrix according to the local grain boundary velocity. The growth of a nucleus stops, when it impinges upon other nuclei.

The driving force for recrystallization is considerably affected by recovery. This physical ingredient is incorporated into the simulation by a reduction of the dislocation density in all non-recrystallized cells of the grid. The time law of this reduction is assumed to be an exponential decay with time. If the dislocation density within a cell reaches a critically low value, which is necessary for a further propagation of the grain boundaries, the cell is identified as being recovered. The simulation is terminated when all cells within the grid are either recrystallized or recovered.

To properly account for the textural evolution in aluminium alloys, two mechanisms were incorporated into the model. One being oriented nucleation, the other growth selection. In commercial aluminium alloys three types of nucleation sites, namely grain boundaries, cube bands and large second phase particles, can be distinguished [2-4]. The nuclei stemming from each of these sites show a distinct spectrum of crystallographic orientations. The incorporation of the different nucleus populations into the model follows an approach published by Vatne et.al. [5] for hot rolled aluminium.

Nuclei that form at grain boundaries show mainly crystallographic orientations close to those present in the deformed material [6]. However, due to the large orientation gradients near the grain boundaries the orientation distribution of these nuclei exhibits a larger scatter than the deformation texture. To incorporate this scatter into the model, nuclei with orientations found in the deformation texture were interspersed with random oriented nuclei.

Nuclei in the cube bands, band like structures in the deformed state [4], show crystallographic orientations near the cube orientation. If large hard particles are present in the alloy, the deformation zones around these particles, having high orientation gradients and stored energy, form a preferred nucleation site [7]. For the purpose of this model the orientation distribution of these nuclei was assumed to be random.

In most metals, the grain boundary mobility shows a strong dependence on the misorientation between the adjacent grains. During recrystallization, this leads to a preferred growth of nuclei that exhibit crystallographic orientations with high mobilities with respect to their surrounding matrix. In aluminium alloys grain boundaries with a $40°<111>$ orientation relationship ($\Sigma 7$) are known as highly mobile, whereas small angle grain boundaries (SGBD) are generally regarded as immobile. Thus, in the model grain boundaries with orientation differences below $15°$ (SGBD) are assumed to be immobile. Grain boundaries that exhibit the $\Sigma 7$ orientation relationship within $\pm 5.7°$ are assigned a mobility that is 5 times as high as for all other large angle grain boundaries.

RESULTS

The scope of this investigation was to check the model against experimental data. In the following, two examples will be shown. In both cases 2000 nuclei were formed in a grid containing 200^3 individual cells. Nucleation was assumed to occur site saturated.

The first example was an overaged Al-4.5%Cu alloy that was rolled from the as cast state to about 95% thickness reduction at 100°C and then annealed 15s at 540°C. The crystallographic texture of the rolled specimen was discretised using the method devised by Fortunier and Hirsch [8], which considers 936 texture components. This orientation distribution served as input data for the simulation. Fig.1 shows the orientation distribution function (ODF) of this input data. The ODF is displayed in sections perpendicular to φ_2 through Euler space.

Particle stimulated nucleation (PSN) was held solely responsible for the formation of nuclei. Cube bands and grain boundaries were excluded as nucleation sites, since the as cast specimen had a nearly random texture and large grains. The effect of recovery on recrystallization could be neglected in the model because of the short annealing time at a high annealing temperature.

The orientation distribution of the nuclei and the recrystallization texture of the simulation are shown in Fig.2a and Fig.2b respectively. The effect of growth selection can clearly be seen when comparing the nearly random oriented nucleus spectrum with the orientation distribution after recrystallization. The data obtained from this simulation is in good accord with the experiment (Fig.4a) qualitatively as well as quantitatively. It shows a relatively weak ND rotated cube component. The exact position of this texture component, however, differs between simulation and experiment. An explanation for this difference can be found in recent EBSD measurements of nuclei stemming from PSN [9]. These show that the orientation distribution of PSN nuclei is not exactly random. It rather shows a rotation relationship between PSN nuclei and the deformation texture around the <112> axis.

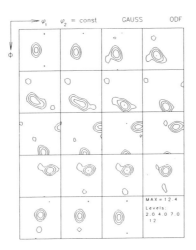

Fig.1: ODF of 95% rolled Al-4.5% Cu

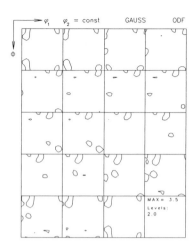

Fig.2:
a) ODF of simulated PSN nuclei.   b) simulated ODF of recrystallized Al-4.5%Cu
                                             assuming PSN and growth selection.

The second example was a commercial purity aluminium alloy (AA1145) rolled to 98% thickness reduction and annealed 20s at 250°C [10]. The ratio of the different nucleus populations was varied until good fit was obtained. The distribution of the nuclei was 80% grain boundary nucleation, 15% nucleation at cube bands, 5% PSN. Results of the calculation are shown in Fig.3. Again the results show good agreement with the experimental data (Fig.4b).

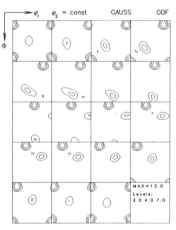

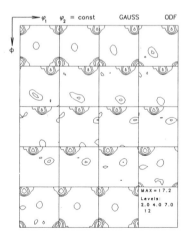

Fig.3:
a) ODF of simulated nuclei in AA1145.   b) simulated ODF of recrystallized AA1145.

110

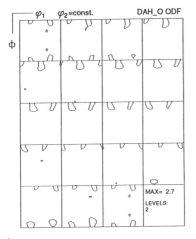
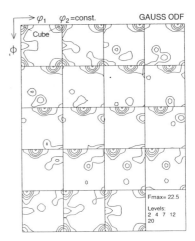

Fig.4:
a) ODF of Al-4.5%Cu rolled 95% at 100°C,   b) ODF of AA1145 rolled 98% at RT,
annealed 15s at 540°C   annealed 25s at 250°C

When comparing the simulated ODF of the nuclei with the simulated ODF after complete recrystallization, a slight shift of the position of the grain boundary nucleated orientations can be noticed. This shift can be attributed to growth selection and is also found experimentally.

For comparison, the ratio of nucleation sites was calculated using an approach developed by Vatne [5]. These calculations yielded the same results as our simulation.

CONCLUSIONS

A model for the simulation of primary static recrystallization in aluminium alloys with modified cellular automata was introduced. It is flexible to incorporate the broad spectrum of nucleation processes that are characteristic for these alloys as well as the growth behaviour of the nuclei. The model allows to simulate crystallographic texture development during primary static recrystallization.

The results of texture simulations on two different aluminium alloys were in good accord with experimental data as well as theoretical predictions.

ACKNOWLEDGEMENTS

This work was supported by the Deutsche Forschungsgemeinschaft (DFG) through the Collaborative Research Centre 370 "Integral Modelling of Materials".

REFERENCES

1. F.R. Reher, doctoral thesis, RWTH Aachen.

2. V. Marx, F.R. Reher, G. Gottstein, submitted to Acta mater.

3. D. Juul Jensen, N. Hansen and F.J. Humphreys, Acta metall. **33**, 2155 (1985).

4. K. Lücke and O. Engler, Mat. Sci. Tech. **6**, 1113 (1990).

5. J. Hjelen, R. Ørsund and E. Nes, Acta metall. mater. **39**, 1377 (1991).

6. H.E. Vatne, T. Furu, R. Ørsund and E. Nes, Acta mater. **44**, 4463 (1996).

7. S.P. Bellier and R.D. Doherty, Acta metall **25**, 521 (1977).

8. F.J. Humphreys, Acta metall. **25**, 1323 (1997).

9. Fortunier and J. Hirsch, in Theoretical Methods of Texture Analysis, edited by H.J. Bunge (DGM 1987), p. 231.

10. O. Engler, X.W. Kong and P. Yang, Scripta mater. in press.

11. O. Engler, habilitation thesis, RWTH Aachen (1995).

A FINITE ELEMENT METHOD FOR SIMULATING INTERFACE MOTION

H.H. YU and Z. SUO
Department of Mechanical and Aerospace Engineering, Princeton Materials Institute
Princeton University, Princeton, NJ 08544

ABSTRACT

This paper describes our recent progress in developing a finite element method for simulating interface motion. Attention is focused on two mass transport mechanisms: interface migration and surface diffusion. A classical theory states that, for interface migration, the local normal velocity of an interface is proportional to the free energy reduction associated with a unit volume of atoms detach from one side of the interface and attach to the other side. We express this theory into a weak statement, in which the normal velocity and any arbitrary virtual motion of the interface relate to the free energy change associated with the virtual motion. An example with two degrees of freedom shows how the weak statement works. For a general case, we divide the interface into many elements, and use the positions of the nodes as the generalized coordinates. The variations of the free energy associated with the variations of the nodal positions define the generalized forces. The weak statement connects the velocity components at all the nodes to the generalized forces. A symmetric, positive-definite matrix appears, which we call the viscosity matrix. A set of nonlinear ordinary differential equations evolve the nodal positions. We then treat combined surface diffusion and evaporation-condensation in a similar method with generalized coordinates including both nodal positions and mass fluxes. Three numerical examples are included. The first example shows the capability of the method in dealing with anisotropic surface energy. The second example is pore-grain boundary separation in the final stage of ceramic sintering. The third example relates to the process of mass reflow in VLSI fabrication.

INTRODUCTION

When a solid is held at an elevated temperature for some time, its structure changes. A film may break into droplets, and an interconnect may grow cavities. The changes are brought about by atomic movements such as diffusion, and motivated by the reduction of free energy such as surface energy. For a bulk material, a detailed knowledge of its microstructure is less critical because an overall knowledge, such as the grain size distribution and pore volume fraction, is often adequate. For a film or a line, where the grain size is comparable to the film thickness and line width, an overall knowledge of structure is inadequate; for example, in submicron aluminum interconnects, the electromigration damage relates to structural details, e.g., crystalline texture and individual grain-boundary orientation [1]. The feature size in integrated electronic circuits is now close to 100 nanometers. For such a small structure, the details of the structure and their evolution are essential to its performance. We can now analyze deformation in complex structures using finite element computer codes. Can we develop a similar tool for evolving structures? This paper describes the recent development of an approach that treats surface motion in a way that resembles the finite element analysis of deformation. The approach naturally combines multiple mass transport mechanisms and energetic forces. For clarity we will introduce the method for interface migration first and then for combined surface diffusion and evaporation-condensation. Surface diffusion alone can be treated as a liming case by setting a very small evaporation-condensation rate.

INTERFACE MIGRATION

Weak Statement

Our method is built upon a weak statement of the interface migration problem. Here we outline its main features; more detailed discussion are given in [2]. Consider an interface separating either two phases of the same atomic composition, or two grains of the same crystalline structure. The interface migrates when the material on one side grows at the expense of the material on the other side. The motion is driven by the reduction of total free energy associated with the motion. Let γ be the surface energy density, which may depend on crystalline orientation. Let g be the difference in the free energy density of the two phases. In evaporation-condensation it is the free energy increase associated with the condensation of unit volume of solid. The total free energy is

$$G = \gamma A + gV,\qquad(1)$$

where A is the area of the interface and V the volume of the solid phase. If γ is anisotropic, the total surface energy is a sum over all the facets, or an integral over the entire interface. Thermodynamics requires that the reaction proceed to decrease the free energy.

Define a quantity p such that

$$\int p\,\delta r_n\,dA = -\delta G,\qquad(2)$$

where δr_n is the virtual motion of the interface, and δG the free energy change associated with the virtual motion. The virtual motion is a small movement in the direction normal to the interface that need not obey any kinetic law. It is an arbitrary function of the position of the interface. Consequently, equation (2) defines the quantity p at every point on the interface. It has the unit of pressure and is the free energy reduction per unit interface area moving per unit distance.

The interface migrates as atoms leave one side and attach to the other side. The actual normal velocity of the interface is a function of the driving pressure. We adopt the linear kinetic relation:

$$v_n = mp.\qquad(3)$$

Here m is the mobility of the interface, which is a given quantity in the simulation.

Eliminating p from (2) and (3), we have

$$\int \frac{v_n}{m}\delta r_n\,dA = -\delta G.\qquad(4)$$

Equation (4) holds for any distribution of the virtual motion, δr_n. We refer to (4) as the *weak statement* of the interface migration problem. It has a weaker requirement on the smoothness of the interface than that of the conventional method described by a set of partial differential equations. When the surface tension γ is isotropic, the weak statement is equivalent to the curvature driven migration. When γ is anisotropic, the curvature driven migration is incorrect, but the weak statement remains valid. The following example shows how this weak statement works.

An Example

Figure 1 shows a crystal having anisotropic surface tension such that it grows into a prism with a square cross-section. When a small particle of such a crystal is introduced into its vapor, it has two degrees of freedom: both the base side B and the height C can change. We need to determine the side and height as functions of time, $B(t)$ and $C(t)$. The surface tension on the prism bases and sides are γ_1 and γ_2, and the mobilities are m_1 and m_2. When the crystal grows by a unit volume at the expense of the vapor, the phase change alone increases the free energy by g. In this discussion, we assume that g is a constant independent of B and C. The total free energy of the system, relative to the vapor without the particle, is

$$G(B,C) = 2\gamma_1 B^2 + 4\gamma_2 BC + gB^2C .\qquad(5)$$

Associated with the virtual changes δB and δC, the free energy varies by

$$\delta G = (4\gamma_1 B + 4\gamma_2 C + 2gBC)\delta B + (4\gamma_2 B + gB^2)\delta C .\qquad(6)$$

The kinetic term on the left-hand side of (4) is

$$\int \frac{v_n}{m}\delta r_n\, dA = \frac{\dot{B}BC}{m_2}\delta B + \frac{\dot{C}B^2}{2m_1}\delta C .\qquad(7)$$

The superimposed dot stands for a time derivative. The weak statement (4) requires that both the coefficients before δB and δC vanish, giving

$$\dot{B} = -m_2\left(\frac{4\gamma_1}{C} + \frac{4\gamma_2}{B} + 2g\right), \quad \dot{C} = -m_1\left(\frac{8\gamma_2}{B} + 2g\right)\qquad(8)$$

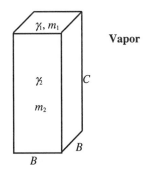

Figure 1. A crystal having anisotropic surface tension grows in its vapor to a prism with a square cross-section.

These are coupled nonlinear ordinary differential equations for the two functions $B(t)$ and $C(t)$, which can be integrated numerically once the initial particle dimensions are given.

Finite Element Method

Equation (4) can be solved numerically by finite element method. We have developed a finite element program for simulating evaporation-condensation [3]. The general procedure is similar to the one we just illustrated in the above example. We approximate an interface by many small elements. The interface is then represented by the positions of all the nodes which are the generalized coordinates $q_1, q_2, q_3, ..., q_{3n-2}, q_{3n-1}, q_{3n}$, where n is the total number of the nodes. The generalized velocities are $\dot{q}_1, \dot{q}_2, \dot{q}_3, ..., \dot{q}_{3n-2}, \dot{q}_{3n-1}, \dot{q}_{3n}$ and the virtual motion of the interface is represented by $\delta q_1, \delta q_2, \delta q_3, ..., \delta q_{3n-2}, \delta q_{3n-1}, \delta q_{3n}$. The velocity and virtual motion of a point on the interface can be interpolated by the values of the nodes. Do the integration in the weak statement (4) element by element, we get a bilinear form in $\dot{q}$ and δq similar to (7). The right-hand side of (4) is the total free energy change associated with the virtual motion,

$$\delta G = -\sum f_i \delta q_i, \qquad (9)$$

which allows us to compute the generalized forces $f_1, f_2, ..., f_{3n}$. Collect the coefficient of δq_i, giving

$$\sum_j H_{ij} \dot{q}_j = f_i. \qquad (10)$$

Equation (10) is a set of linear algebraic equations for the velocity components. Once solved, they update the nodal positions for a small time step. The process is repeated for many steps to evolve the interface. Because the matrix H and force column f depend on the coordinates of all the nodes q, (10) is a nonlinear dynamical system. H is a symmetric and positive-definite matrix, which we call the viscosity matrix.

Numerical Example

Figure 2 shows a numerical example of a grain growing in a vapor. The surface energy is anisotropic with a four-fold symmetry,

$$\gamma(\theta) = \gamma_0 \left(1 + |\theta|\right) \qquad \text{for } |\theta| \leq \pi/4.$$

Figure 3 is the polar plot of the surface energy density. The dash-dot line represents the corresponding Wulff shape. In Fig. 2, we start with a small particle with the surfaces having high surface energy density. The particle is immersed in the vapor which has a higher phase energy density than the particle. The normalized phase energy density difference between the particle and the vapor is $g a_0 / \gamma_0 = 3.0$, with a_0 being the half diagonal length of the initial particle. The particle grows and, at the same time, forms small facets with the lowest surface energy density. These small facets finally disappear and the final shape of the particle is the same as the Wulff shape in Fig. 3. The size of the small facets in the simulation is not physical; it depends on the size of the finite elements. The trend in the simulation should be correct.

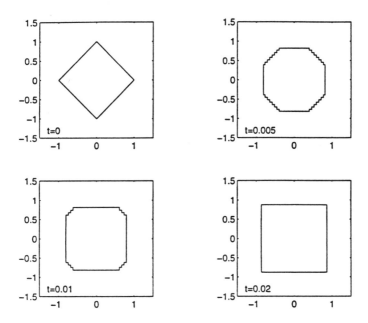

Figure 2. Anisotropic crystal growth in a vapor. The surface energy depends on surface orientation and has a four-fold symmetry. The particle evolves to the Wulff shape.

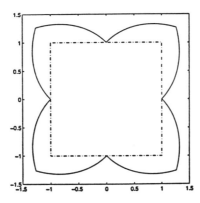

Figure 3. Polar plot of the surface energy density of the particle. The Wulff shape is a square.

The simulation should have captured the general feature of anisotropic grain growth. The unit of time in Fig. 2 is $a_0^2/m\gamma_0$, with m being the mobility of the surface.

SURFACE DIFFUSION AND EVAPORATION-CONDENSATION

Weak Statement

Sun and Suo [4] formulated the weak statement for combined surface diffusion and evaporation-condensation. Imagine two concomitant processes on a surface: the solid matter can relocate on the surface by diffusion, and exchange with the surrounding vapor by evaporation-condensation. The combination of these two processes reduces the total free energy of the system. The relocation of the atoms on the surface and the mass exchange between solid and vapor both change the geometry of the surface. For a polycrystalline particle, its surface intersects with some grain boundaries. The geometric change of the surface will also change the total grain boundary area. So the total free energy change should include the change due to energy density difference between solid and vapor, the change of interfacial energy due to area, and the change of the orientation of the surface and grain boundaries.

Figure 4 illustrates a surface in three dimensions. Denote the unit vector normal to the surface element by n. An arbitrary contour lies on the surface, with the curve element dl, and the unit vector m in the surface and normal to the curve element. At a point on the contour, m and n are perpendicular to each other, and both are perpendicular to the tangent vector of the curve at the point. Mass flux J is a vector field tangent to the surface and is defined such that $J \cdot m$ is the volume of atoms crossing unit length of the curve in unit time. The mass exchange between the solid and vapor is represented by flux j, the volume of matter added to a unit area of the solid surface per unit time due to evaporation-condensation. Both the mass flux J due to surface diffusion and the mass flux j from the vapor to the solid change the geometry of the surface. Mass conservation relates the surface velocity, v_n, to the fluxes of the two matter transport processes:

$$v_n = j - \nabla \cdot J . \tag{11}$$

If we neglect the surface diffusion part in (11), j is the normal velocity of the surface. This is the case we considered in the last section for interface migration alone.

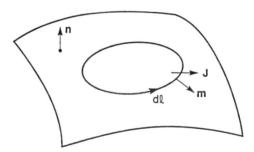

Figure 4. A surface in three dimensions.

The virtual motion due to surface diffusion can be represented by a vector field δI tangent to the surface, defined such that $\delta I \cdot m$ is the volume of atoms crossing a unit length of the curve. The virtual motion due to evaporation-condensation can be expressed by δi, the volume of matter added to a unit area of the solid. The two kinds of virtual mass fluxes give the surface a virtual normal displacement,

$$\delta r_n = \delta i - \nabla \cdot (\delta I).$$ (12)

Associated with the virtual motion, the free energy changes by δG. Define both the driving pressure p for evaporation-condensation and the driving force F for surface diffusion in one integral form,

$$\int (F \cdot \delta I + p \delta i) dA = -\delta G.$$ (13)

The integral extends over all the surface areas participating in mass transfer. The equation holds for any virtual motion. Because δI is arbitrary, the local quantity F is prescribed by this global statement. This definition is equivalent to Herring's [5] definition in which the driving force for surface diffusion, F, is the amount of free energy decrease associated with per unit volume of matter moving per unit distance on the surface. F is also a vector field tangent to the surface. When the surface tension is isotropic, the definition in (13) can reproduce the relation between the driving force and the curvature gradient [5]. The weak statement is valid for anisotropic surface tension. The weak statement also enforces the local equilibrium condition at the triple junction where three interfaces meet.

Following Herring, we write a kinetic law that, at every point on the surface, the flux is proportional to the driving force:

$$J = MF.$$ (14)

Here M is the mobility of atoms for surface diffusion. We assume that the mobility is isotropic. It is represented by a number M, related to the self-diffusivity by the Einstein relation, $M = \Omega D \delta / kT$, where Ω is the volume per atom, D the self-diffusivity on the surface, δ the effective thickness of atoms participating in matter transport, k Boltzmann's constant, and T the absolute temperature.

The kinetic law for evaporation-condensation is the same as (3),

$$j = mp,$$ (15)

where m is the specific rate of evaporation-condensation.

Substituting (14) and (15) into (13), we obtain that

$$\int \left(\frac{J \delta I}{M} + \frac{j \delta i}{m} \right) dA = -\delta G.$$ (16)

This is the *weak statement* for the combined evaporation-condensation and surface diffusion.

In formulating a finite element method, it is convenient to use δI and δr_n as the basic variables. Eliminating j and δi in (16) by using (11) and (12), we obtain that

$$\int \left\{ \frac{J \cdot \delta I}{M} + \frac{(v_n + \nabla \cdot J)[\delta r_n + \nabla \cdot (\delta I)]}{m} \right\} dA = -\delta G. \tag{17}$$

In this form, the weak statement only involves two virtual fields, δr_n and δI. They vary independently, subject to no constraint. The free energy change on the right-hand side of (17) includes the energy change associated with the phase transformation and the interfacial energy change associated with the area and orientation change of the surface and grain boundaries.

Surface diffusion alone can be treated as a limiting case. In this situation, matter only diffuses on the surface but does not exchange between the solid and the vapor. Thus we set a small value to m in (17). In this case, the free energy change on the right-hand side of (17) associated with the phase transformation can be neglected. Only the surface energy and grain boundary energy need be considered. If λ is a representative length in a problem, the dimensionless parameter $m\lambda^2 / M$, measures the relative rate of evaporation-condensation and surface diffusion. In the numerical simulation, we set $m\lambda^2 / M = 10^{-6}$.

<u>Finite Element Implementation</u>

Same as the method for interface migration, we can solve J and v_n in (17) numerically by finite element method [4]. We can divide the surface into many small elements, and interpolate J and v_n on each element by their values at the nodes of the element. The generalized coordinates include both the positions and the fluxes of the nodes. Evaluate both the left-hand side and the right-hand side of (17) and collect the coefficients of the virtual motion coordinates,

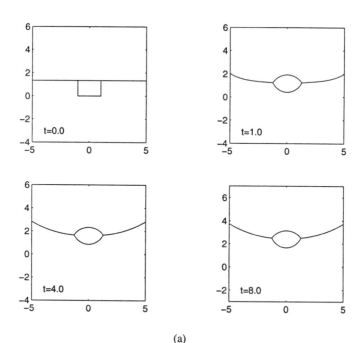

(a)

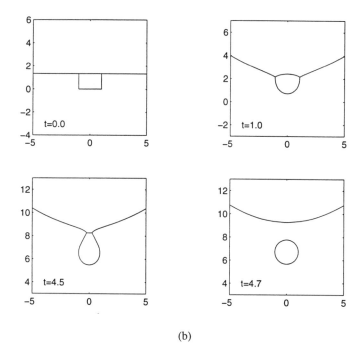

(b)

Figure 5. Pore-grain boundary interaction. A grain boundary is dragged by a force f_b exerted at the outside edge of the simulating unit. The grain boundary migrates. The pore accommodates its position by surface diffusion. In both simulations, we choose $\gamma_B/\gamma_S = 1.0$, $f_b/\gamma_S = 0.5$ and $R/r_0 = 5.0$. a) Pore-grain boundary attachment, $mr_0^2/M = 1.0$; b) Pore-grain boundary separation, $mr_0^2/M = 10$.

and we obtain a set of linear algebraic equations for the velocity and flux components of all the nodes. Solving the equations, we can update the nodal positions for a small time step. The process is repeated for many steps to evolve the surface.

Pore-Grain Boundary Separation

Yu and Suo [6] developed a finite element program for simulating three dimensional axisymmetrical microstructure evolution. The program was used to simulate the process of pore-grain boundary separation. In the final stage of ceramic sintering, the individual pores are isolated on grain boundaries. At this stage, some grains may grow by grain boundary motion. A pore on a migrating grain boundary may either migrate with, or break away from, the grain boundary. The pore accommodates its position by evaporation-condensation, surface diffusion, and volume diffusion. For a small pore, surface diffusion dominates [7,8]. In our simulation, we just considered the surface diffusion of the pore and the grain boundary migration. We

121

focused on the process of separation, and neglected the grain boundary diffusion in our simulation.

Figure 5 shows two examples. In our simulation, we prescribed a force f_b at the outside edge (a circle) of the simulation unit. This force may represent the force exerted by other grain boundaries intersecting with the migrating grain boundary. The dimensionless parameters are the ratio of grain boundary energy and surface energy γ_B/γ_s ; the normalized prescribed force f_b/γ_s ; the relative rate of grain boundary migration and surface diffusion mr_0^2/M , with m being the mobility of the grain boundary, M the mobility of the surface diffusion and r_0 the radius of a sphere which has the same volume as the pore; the ratio of the radius of the simulation unit and the pore size R/r_0 . R can be considered as the grain size. In our simulation, we took the initial pore shape to be a cylinder. In Fig. 5a, the pore changes the initial shape and finally moves with grain boundary at a constant velocity. The angles at the triple junction quickly evolve to the equilibrium dihedral angles. This local equilibrium condition is enforced by the weak statement naturally. In Fig. 5b, with all other parameters held the same as in Fig. 5a, we changed the relative mobility, mr_0^2/M . In this case the surface diffusion of the pore is so slow compared with the grain boundary migration that the pore cannot accommodate its position quickly enough to catch up with the movement of the grain boundary. Finally the pore separates from the grain boundary. Once trapped inside the grain, the pore heals into a sphere. The unit of time is $r_0^4/M\gamma_s$, and all the lengths are normalized by r_0.

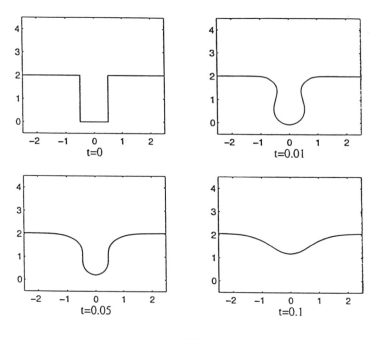

(a)

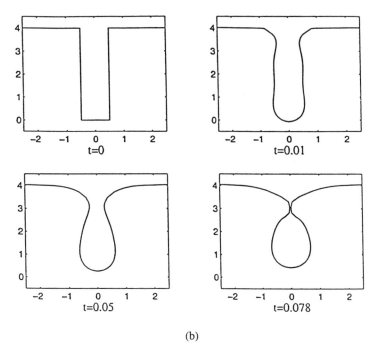

(b)

Figure 6. Mass reflow in a cylindrical hole. a) Mass fills the hole with aspect ratio $d/h = 1/2$;
b) A void forms in the hole with aspect ratio $d/h = 1/4$.

Mass Reflow

Mass reflow on a free surface with deep trenches and vias is used in fabricating metal interconnects. After a metal thin film is deposited on a substrate at a low temperature, the trenches and vias on the substrate are not fully filled with the metal. The metal film is then heated to some temperature so that mass can diffuse on the surface. Mass on the platforms can flow into the trenches and vias, which result in the desired interconnects. However, sometimes voids may form before mass fills in the trenches and vias [9]. Gardner and Fraser [10] simulated the effect of trench dimension and film thickness on the mass flow. Huang, Gilmer and de la Rubia [11] developed an atomistic simulator for thin film deposition in three dimensions and simulated the mass reflow of Al thin film. Sun [12] simulated the mass reflow both in a trench and a hole by using the finite element method described in this paper. Figure 6 is the simulation for mass reflow in an axisymmetric hole. In Fig. 6a, the aspect ratio of the hole diameter over the hole depth is $d/h = 1/2$. The hole can be filled without the formation of void. In Fig. 6b, the aspect ratio of the hole diameter over the hole depth is $1/4$, the void is formed before the hole is filled completely. In Fig. 6, the unit time is $d^4/M\gamma_s$. All the lengths are normalized by d.

123

SUMMARY

We have described the finite element method for simulating interface motion due to interface migration and surface diffusion. The method is based on a weak statement which has a weaker requirement on the smoothness of simulating surface than that of the conventional motion by curvature method. The method can handle surface energy anisotropy. The triple junction at which three interfaces meet automatically evolves to the equilibrium dihedral angles. Numerical examples demonstrate that the method can capture the intricate details in transient motions. The method can readily include multiple energetic forces and rate processes, and is applicable to diverse problems with large geometry change.

ACKNOWLEDGEMENTS

The work is supported by the National Science Foundation through a Young Investigator Award and by the Institute for Materials Research and Engineering, Singapore.

REFERENCES

1. C.V. Thompson and J.R. Lloyd, MRS Bulletin, December 1993, p. 19 (1993).
2. Z. Suo, Advances in Applied Mechanics **33**, p.194 (1997)
3. B. Sun, Z. Suo and W. Yang, Acta Mater. **45**, p.1907 (1997).
4. B. Sun and Z. Suo, Acta Mater. **45**, p. 4953 (1997).
5. C. Herring, in The Physics of Powder Metallurgy, edited by W.E. Kingston, pp. 143-179, McGraw-Hill, New York (1951).
6. H.H. Yu and Z. Suo, unpublished work.
7. P.G. Shewmon, Trans. Met. Soc. AIME **230**, p.1134 (1964).
8. C.H. Hsueh and A.G. Evans, Acta Metall. **31**, p.189 (1983).
9. Y. Arita, N. Awaya, K. Ohno and M. Sato, MRS Bulletin **19**, p.66 (1994).
10. D.S. Gardner and D.B. Fraser, Presentation in MRS conference, spring, 1995.
11. H. Huang, G.H. Gilmer and T. D. de la Rubia, An atomic simulator for thin film deposition in three dimensions, preprint (1997).
12. B. Sun, Ph.D. thesis, University of California at Santa Barbara, p.39 (1996).

THREE DIMENSIONAL SIMULATION OF THE MORPHOLOGICAL EVOLUTION OF A STRAINED FILM ON A THICK SUBSTRATE

Cheng-hsin Chiu, Institute of Materials Research and Engineering, Blk S7 Level 3, National University of Singapore, Singapore 119260.

ABSTRACT

This paper presents a three-dimensional simulation for the surface evolution of a strained film on a thick substrate. The simulation shows that an initially random morphology will first transform into a profile dominated by two-dimensional ridges, and then into a three-dimensional island surface. The simulation also demonstrates that the ridge-island transition is a kinetic process which can be delayed by changing the initial morphology.

INTRODUCTION

It is well known that the morphology of a heteroepitaxial thin film on a thick substrate may develop a rough or an island profile during growth and/or upon subsequent annealing [1–3]. The surface roughening process, on one hand, may induce defects on the film surface; thus, this process has to be suppressed or delayed in some applications [2, 4]. On the other hand, the surface roughening process can be utilised to fabricate island/dot structures for the manufacture of self-assembled quantum dots [5]. In both cases, it is important to understand the roughening process in order to control the morphological evolution and to obtain the desired morphology.

Because of this importance, the surface evolution of this type of structure has been investigated intensively from both experimental and theoretical points of view. The theoretical analyses include the stability of a flat film surface against perturbations [6–14], equilibrium surface shapes [15, 16], and simulations of surface evolution [17–25]. Those analyses make significant contributions toward comprehending the morphological evolution of the structure. However, most of the analyses are two-dimensional; as a consequence, the results cannot explain experimental findings which involve three-dimensional configurations, for example, the island-ridge-island transition [2, 3].

In order to overcome the limit in analysis, we are currently developing physical models and simulation techniques for investigating the three-dimensional case of the morphological evolution of heteroepitaxial film/substrate structures. In this paper, we report our recent progress in studying the case where the film is thick, the material properties of the system are isotropic, and the evolution is dominated by surface diffusion.

PROBLEM STATEMENT

Consider a film coherently bonded to a semi-infinite substrate. A set of Cartesian coordinate axes is attached to the neutral plane of the film surface with the x- and the y-axes being placed parallel to the plane and the z-axis normal to the plane. The film surface profile is assumed to be periodic in the x- and the y-directions with the same wavelength λ; in other

125

words, the surface is composed of identical square cells. The morphology of the square cell at time t can be described by a Fourier series as

$$f(x, y, t) = \text{Re} \left\{ \sum_{m,n=0}^{\infty} a_{mn}(t) e^{i2\pi(mx+ny)/\lambda} + a_{m\bar{n}}(t) e^{i2\pi(mx-ny)/\lambda} \right\}. \tag{1}$$

For convenience, the set of the Fourier coefficients in eqn (1) is denoted as $\mathbf{A}(t)$.

In contrast to the rough film surface, the interface between the film and the substrate is flat and remains at $z = -h$ for all times where h is the film thickness. It is assumed that h is much larger than the amplitude of the wavy film surface.

The film and the substrate are isotropic materials and they have the same shear modulus μ and Poisson ratio ν. The lattice sizes of the two materials, however, are different, causing stresses in the structure. The stresses can be determined by solving the elasticity problem where there is mismatch strain ϵ^T in the film and the traction vanishes on the film surface [11, 26]. The mismatch strain ϵ^T is given by $\epsilon^T \mathbf{I}$ where $\epsilon^T = (a_f - a_s)/a_s$.

We assume the morphological evolution of the film surface is controlled by surface diffusion, which can be expressed as [6, 27],

$$\frac{df(x, y, t)}{dt} = \frac{\Omega^2}{n_z} \mathbf{\nabla}^\Gamma \cdot \left[\frac{\rho_s D_s}{k_B T_k} \mathbf{\nabla}^\Gamma (w - \gamma\kappa) \right]. \tag{2}$$

In eqn (2), Ω is the atomic volume, n_z is the z-component of the normal vector $\mathbf{n}$ of the film surface, $\mathbf{\nabla}^\Gamma$ is the surface gradient operator [28], ρ_s is the adatom density, D_s is the surface diffusivity, k_B is the Boltzmann constant, T_k is the temperature, w is the strain energy density, γ is the surface energy density, and κ is the curvature. Equation (2) and the elasticity problem mentioned in the last paragraph constitute the surface evolution problem to be solved in this paper. For convenience, the solution to the problem is normalised by the following characteristic energy density, length, and time

$$w_0 = \frac{(1-\nu)T^2}{4\mu}, \quad L = \frac{\gamma}{w_0}, \quad t_L = \frac{k_B T_k \gamma^3}{\rho_s D_s \Omega^2 w_0^4} \tag{3}$$

where $T = -2\mu(1+\nu)\epsilon^T/(1-\nu)$. Typical values of L and t_L of the SiGe system can be found in [19]. It is noted that the value $1.5w_0$ is the strain energy density on a flat film surface when $\nu = 1/3$ [25].

NUMERICAL ANALYSIS

The solution procedure for determining the evolution of the film surface can be briefly described as follows. (I) We solve the three-dimensional elasticity problem and calculate the strain energy density on the film surface by a high-order boundary perturbation method [29]. (II) We use the fast Fourier transform (FFT) technique to evaluate the curvature κ and the surface gradient operator $\mathbf{\nabla}^\Gamma$. (III) We then substitute w, κ, and $\mathbf{\nabla}^\Gamma$ into eqn (2), and integrate the result with respect to time by the mid-point rule to derive $f(x, y, t)$. The procedure is an improvement on those in [19, 24, 25]. Particularly, the current procedure adopts the boundary perturbation method as the elasticity solver, which is an efficient and accurate scheme for obtaining the elasticity solutions. The new solver is the key to accomplishing the three-dimensional simulation presented in this paper.

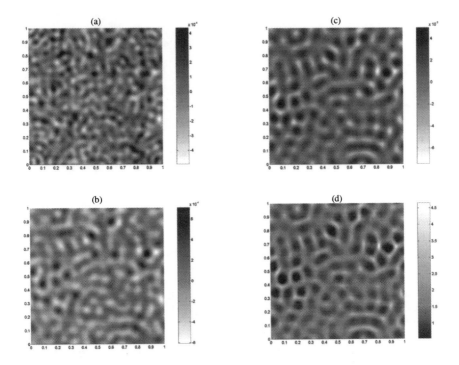

Figure 1: The film surfaces $f(x, y, t)/\lambda$ at (a) $t/t_L = 0$, (b) 0.0249, and (c) 0.1243 for the case where the initial roughness is random and small, and (d) the strain energy density w/w_0 on the surface shown in Part (c)

SIMULATION RESULTS

In this section, we employ our numerical method to study the three-dimensional cases of the surface diffusion-controlled morphological evolution. We take ν to be $1/3$, and we focus our attention on how the evolution is affected by the initial conditions.

The initial morphologies of the surface cells considered in this paper are depicted in Part (a) of Figures 1–3. The size λ of the cells is chosen to be $20\pi L/3$, which is ten times the wavelength of the fastest surface diffusion mode [6, 7]. The size λ is about $5\mu m$ for the $Si_{1-x}Ge_x$ system with x being 0.165 [19].

The initial surface profiles are obtained as follows. We first use a random-number generator to create a set of Fourier coefficients $\mathbf{A}'$. We then substitute the set $\mathbf{A}'$ into the first-order perturbation solution of surface diffusion [6, 7] to calculated the Fourier coefficients when the fastest diffusion modes in $\mathbf{A}'$ increase by 8%. The resultant coefficient set, denoted as $\mathbf{A}_0$, produces the random roughness shown in Part (a) of Figure 1. Reducing the components $\{a_{mn}, a_{m\bar{n}} | n \neq 0\}$ in $\mathbf{A}_0$ to 5% of their original values generates the profile in Figure 2 (a), which is characterised by ridges lying in the y-axis. Modifying the components $\{a_{mn}, a_{m\bar{n}} | \tan^{-1}(n/m) > 15°\}$ of $\mathbf{A}_0$ in a similar way yields the morphology in Figure 3 (a).

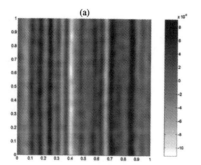 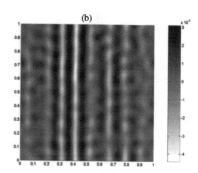

Figure 2: The film surfaces $f(x, y, t)/\lambda$ at (a) $t/t_L = 0$ and (b) 0.1789 for the case where the initial profile is dominated by ridges lying along the y-axis.

Figure 1 plots the morphological evolution which starts from a surface with random roughness: Parts (a)-(c) depict the surface profiles $f(x, y, t)/\lambda$ during the evolution and Part (d) depicts the strain energy density w/w_0 on the surface shown in (c). The surface profiles are illustrated by a grey scale with black corresponding to the highest peaks and white to the deepest valleys. The results show that the amplitude of the rough surface increases, and the characteristics of the morphology experience a two-stage transition during the evolution. The morphology first transforms into a configuration dictated by ridges, and then evolves to an island surface as the ridge-island transition occurs. Besides the islands, sharp valleys also develop during the surface roughening process. The length of the sharp valleys is approximately equal to the size of the adjacent islands.

Figure 2 depicts the evolution of long straight ridges where Part (a) is the initial condition and (b) is the surface profile at $t/t_L = 0.1789$. It is found that at $t/t_L = 0.1789$ the ridge-island transition has occurred along the slightly raised ridges but not along the elevated ones. Sharp valleys also develop in this case. However, the sharp valleys are much longer than those in Figure 1, and they are located between the elevated ridges.

Figure 3 presents the third example where the initial profile is characterised by ridges with the orientation θ in the range $[-15°, 15°]$ where θ is measured from the y-axis. In this example, the formation of round islands is almost suppressed, leading to a ridge-dominated morphology throughout the evolution. The surface sites between the tall ridges are favourable locations for the development of sharp valleys; the size of the valleys is between those in the first two cases.

It is interesting to note that though the morphologies in the three examples are different, the evolutions of the spacing between the ridges/islands in those examples are similar. The spacing is small initially; it then increases toward the wavelength of the fastest surface diffusion mode given by $2\pi L/3$.

DISCUSSION

It is necessary to explain the occurrence of the ridge-dominated surface and the transition from ridges to islands indicated in Figure 1. The occurrence of the ridge-dominated surface

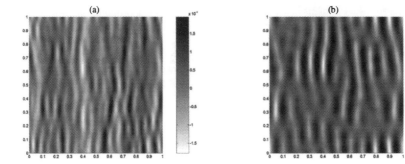

Figure 3: The film surfaces at (a) $t/t_L = 0$ and (b) 0.1507 for the case where the initial profile is dominated by ridges with the orientation in the range $[-15°, 15°]$.

is caused by the initial random roughness. The Fourier coefficients of this type of roughness are characterised by a uniform distribution of magnitude for a wide range of wavelengths. As the evolution proceeds, the Fourier coefficients at large wavelengths become much smaller than those at moderate wavelengths since the former increase at much slower rates. The coefficients at small wavelengths also become smaller because the surface configurations of small wavelengths are not energetically favourable and will decay quickly. This leads to a relatively uniform distribution of Fourier coefficients at moderate wavelengths, and the corresponding morphology is a ridge-dominated surface.

Turn to the ridge-island transition. The transition involves two mechanisms, namely the growth of the ridges and the formation of islands along the ridges. Both mechanisms are energetically favourable [30], and it is the relative rate of the two mechanisms that determines the transition. If the island formation rate is faster, the transition will occur; in contrast, if the ridge growth rate is faster, the transition will be suppressed. The rates of the two mechanisms are controlled by the amplitude a_{island} of the three-dimensional perturbation along the ridges and the amplitude a_{ridge} of the ridges, respectively [6, 7]. This explains the finding that the ridges in the first example collapse more quickly than those in the other examples, by noting the ratio a_{ridge}/a_{island} in the first example is the smallest. On the other hand, the ratio a_{ridge}/a_{island} in the third example is the largest; accordingly, the ridges in that case have the highest resistance to three-dimensional roughening.

The evolution of the spacing between the islands and the ridges, like the ridge-island transition, is a kinetic process which may vary under different conditions. In this paper, the average spacing evolves toward the value $2\pi L/3$ because the morphological evolution is controlled by surface diffusion. The characteristic spacing will evolve toward a different value if the surface evolution is controlled by other mechanisms such as evaporation/condensation. In the case where the substrate becomes partially exposed or other defects are generated during the evolution, the characteristic spacing may continuously increase, not approaching any particular value [3].

SUMMARY

We present three-dimensional simulation results for the morphological evolution of a strained film on a substrate. The results show that a morphology with random roughness will first transform into a profile dominated by ridges, and then into a three-dimensional island surface. It is found that the ridge-island transition can be delayed by varying the initial surface profile. This suggests the ridge-island transition is a kinetic process controlled by the growth of the ridges and the formation of islands along the ridges

ACKNOWLEDGEMENT—The author is grateful to Professor H. Gao, Professor L. B. Freund, Dr. C. S. Ozkan, and Dr. E. H. Chason for useful discussions.

REFERENCE

[1] D. J. Eaglesham and M. Cerullo, *Phys. Rev. Lett.* **64**, 1943 (1990).

[2] A. G. Cullis, *MRS Bulletin* **21**, 21 (1996).

[3] C. S. Ozkan, Ph.D. dissertation, Stanford University (1996).

[4] J. Tersoff and F. K. LeGoues, *Phys. Rev. Lett.* **72**, 3570 (1994).

[5] D. Bimberg, M. Grundmann, and N. N. Ledentsov, *MRS Bulletin* **23** No. 2, 31 (1998).

[6] R. J. Asaro and W. A. Tiller, *Metall. Trans.* **3**, 1789 (1972).

[7] D. J. Srolovitz, *Acta metall.* **37**, 621 (1989).

[8] H. Gao, *Int. J. Solids Structures* **28**, 703 (1991).

[9] B. J. Spencer, P. W. Voorhees, and S. H. Davis, *Phys. Rev. Lett.* **67**, 3696 (1991).

[10] M. A. Grinfeld, *J. Nonlinear Sci.* **3**, 35 (1993).

[11] L. B. Freund and F. Jonsdottir, *J. Mech. Phys. Solids* **41**, 1245 (1993).

[12] J. E. Guyer and P. W. Voorhees, *Phys. Rev. Lett.* **74**, 4031 (1995).

[13] C.-h. Chiu, and L. B. Freund, Morphological instability of a strained film on a compliant substrate. Manuscript in preparation.

[14] C.-h. Chiu and L. B. Freund, In *Thin Films: Stresses and Mechanical Properties VI*, eds. W. W. Gerberich, H. Gao, J.-E. Sundgren, and S. P. Baker, *MRS Sym. Proc.* **436**, 517 (1996).

[15] R. V. Kukta and L. B. Freund, *ibid*, 493 (1996).

[16] B. J. Spencer and J. Tersoff, *Phys. Rev. Lett*, in press.

[17] C.-h. Chiu and H. Gao, *Int. J. Solids Structures* **30**, 2983 (1993).

[18] W. H. Yang and D. J. Srolovitz, *Phys. Rev. Lett.* **71**, 1593 (1993).

[19] C.-h. Chiu and H. Gao, In *Mechanisms of Thin Film Evolution*, eds. S. M. Yalisove, C. V. Thompson, and D. J. Eaglesham, *MRS Symp. Proc.* **317**, 369 (1994).

[20] F. Jonsdottir and L. B. Freund, *ibid*, 309 (1994).

[21] L. B. Freund, *Acta Mech. Sinica* **10**, 16 (1994).

[22] B. J. Spencer and D. I. Meiron, *Acta metall.* **42**, 3629 (1994).

[23] L. B. Freund, *Int. J. Solids Structures* **32**, 911 (1995).

[24] C.-h. Chiu and H. Gao, In *Thin Films: Stresses and Mechanical Properties V*, eds. S. P. Baker, C. A. Ross, P. H. Townsend, C. A. Volkert, and P. Borgesen, *MRS Symp. Proc.* **356**, 33 (1995).

[25] C.-h. Chiu, Ph.D. dissertation, Stanford University (1995).

[26] H. Gao, *J. Mech. Phys. Solids* **42**, 741 (1994).

[27] W. W. Mullins, *J. Appl. Phys.* **28**, 333 (1957).

[28] M. A. Weatherburn, *Differential Geometry of Three Dimensions*, Cambridge University Press, London (1927).

[29] C.-h. Chiu, High-order boundary perturbation analysis for three-dimensional elasticity problems and its application in simulating the surface evolution of a strained film. Manuscript in preparation.

[30] This can be understood from the first-order perturbation result [6–9]; this is also indicated in our simulation that both the ridges and the islands can grow during the evolution.

Part III

Other Applications of Mathematical Modelling

BONDING REGENERATION: THE DRIVING FORCE OF HETERO-EPITAXIAL DIAMOND GRAIN COALESCENCE ON (001) SILICON

R.Q. Zhang*, X. Jiang**, C.L. Jia*** and S.-T. Lee*
*Department of Physics and Materials Science, City University of Hong Kong, Hong Kong
**Fraunhofer-Institut fuer Schicht- und Oberflaechentechnik (FhG-IST), Bienroder Weg 54 E, D-38108 Braunschweig, Germany
***Institut fuer Festkorperforschung, Forschungszentrum Julich GmbH, D-52425 Julich, Germany

ABSTRACT

The grain coalescence phenomenon in the growth of heteroepitaxial diamond film on (001) silicon substrate by microwave plasma chemical vapor deposition was examined by using high-resolution electron microscopy. It was shown that this phenomenon evidently occurs between two diamond grains with a small-angle tilt. The coalescence was completed after some more growth steps following the meeting of such two grains, indicating the difficulty for the lattice matching in grain boundary. By performing simulation of a step-by-step growth of two diamond grains on a (001) silicon substrate with molecular orbital PM3 method, it was shown that the bonding regeneration between the two grains is essential for the coalescence and the coalescence is only possible when the orientation difference between the grains is sufficiently small so as to allow efficient overlap of electron cloud in the grain boundary. This study indicates that single crystal diamond growth may be possible by the current CVD growth techniques via further reduction of the surface roughness to gain a heteroepitaxy with very small grain tilting.

INTRODUCTION

Diamond films have been deposited homo- or hetero-epitaxially, or as textured films on various substrates such as diamond [1], c-BN [2], SiC [3] and silicon (100) [4-12], (111) [6,8,11] and (110) [13]. Despite these successes towards fulfilling their promise as electronic materials [14-16], the diamond films on inexpensive, large single-crystal silicon substrate are still at best polycrystalline and thus are not yet suitable for the desired electronics application, due to the presence of an amorphous layer in grain boundaries. Further efforts are being made to develop single-crystal diamond films. Among the various approaches, the one taking the advantage of the grain coalescence [17,18] is particularly promising. An understanding of the coalescence mechanism would be very useful for enhancing such a phenomenon and for obtaining high quality single-crystal diamond film.

Electron microscopy is a powerful tool for studying hetero-epitaxial diamond films on various substrates and can give clear pictures of interfaces between the diamond films and the substrates or at the grain boundary [19-22]. These results have also provided information of the nucleation and growth of diamond films. The diamond grain coalescence phenomenon has consequently been observed by using the electron microscopy technique [17,18]. Although the study showed clearly that diamond grain coalescence is due to the interaction during the deposition of some layers, more detailed information such as about the driving force of the coalescence and the coalescence procedure during nucleation and growth is desirable by means of theoretical approach. As far as we know, such information is not yet available. These details of the fundamental nucleation and growth mechanism relating to the coalescence phenomenon of hetero-epitaxial

133

diamond films at the molecular level may provide insight into ways to optimize technological parameters in order to enhance such a coalescence effect and to achieve a high-quality film.

By first examining the coalescence with high-resolution electron microscopy (HREM), this study will further reveal the driving force of such a phenomenon and study the mechanism of the coalescence by means of the molecular orbital PM3 approach. Discussion regarding how to enhance the coalescence to achieve high quality diamond film will also be made.

EXPERIMENTAL AND COMPUTATIONAL APPROACHES

The diamond film was fabricated by microwave plasma-assited chemical vapor deposition (MWCVD). [001] heteroepitaxial flim with thickness of about 10 μm was deposited on 2-inch n-type (001) silicon wafer by applying a negative bias potential to the substrate for heterogeneous nucleation of [001]-oriented diamond crystallites, followed by an MWCVD diamond growth process [23]. The grain boundary of the film was analyzed by transition electron microscopy (TEM).

The PM3 parametrization [24] of the MNDO semi-empirical hamiltonian [25] was used in this study throughout the calculations to optimize the geometrical structures and to calculate the electronic density. *Ab initio* or other higher level calculations were not used due to the complexity of computation for the presently treated large system. PM3 theory has been well established in the literature and has been successfully applied to our previous studies of the diamond growth process on silicon substrates [26-28]. In this work, all of the cluster models have been geometrically optimized, i.e. energetically relaxed.

RESULTS AND DISCUSSIONS

The Grain Coalescence

Figure 1 shows a HREM lattice image of a small angle grain boundary between two heteroepitaxial diamond grains. The image was recorded on a JEOL 400EX microscope, with the electron beam parallel to a <110> zone axis. The left hand side diamond grain (DI) was grown in the ideal [001] parallel heteroepitaxial orientation to silicon. The one on the right (DII) has a small angle deviation from the silicon substrate and the grain on the left. It can be seen by an inspection of the angle with marking lines in the image. From the image, a 2° tilt angle around the viewing direction with respect to the substrate and the left-hand side grain DI can be measured near the interface. The letter "A" denotes the start position of such a low angle grain boundary from the interface between the film and the substrate. Both diamond grains exhibit a multiple-twin structure. The grain boundary shown in Fig. 1 is relatively sharp, indicating that it is at an edge-on position and nearly parallel to the (110) plane. These features of grain boundary were taken into account in our theoretical simulations of this study. The most important information obtained from the figure is the disappearance of the grain boundary at the position indicated by the letter "B". This shows that two initial individual grains coalesced and an on-top single crystal was formed. This result coincides with our SEM observation that after the thickness of the film exceeds about one micron some individual islands grow together and the columnar structures are formed. So, the coalescence phenomenon has evidently occurred.

The above-described coalescence phenomenon can be illustrated in Figure 2. The diamond grains of DI and DII are indicated with their representative frames. DI is ideally epitaxial while DII is about 2.5° tilted due to surface stain induced by lattice mismatch and by surface roughness [28]. Due to the lattice residual mismatch and the lattice displacement of the diamond grain resulted from the surface strain, the boundary between the two grains may not suit for growing round-number diamond lattice units. This is possibly modified by the tilting of DII which actually provides chance for the two grains to approach slightly to each other. However, the tilting may also result in lattice displacement between the two diamond grains. The later factor introduces difficulty for the hydrocarbon species to bridge the two diamond grains. Such a bridging can obviously be regarded as the driving force of the coalescence and is the main reason that the coalescence is only possible for diamond grains with very small angle tilting. This point will be further demonstrated in the following. We note that in this schematic diagram the lattice relaxation that may correct the tilt in a certain extent [28] was not considered when tilt is made.

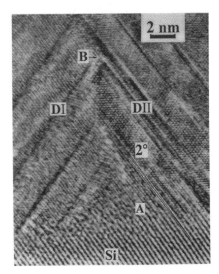

Figure 1. High resolution lattice image of a grain boundary region.

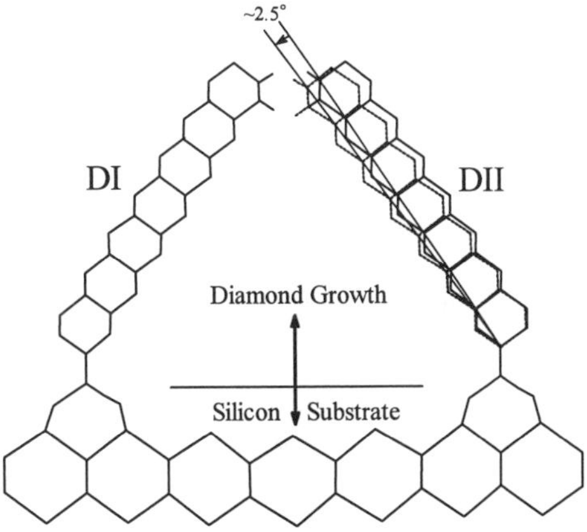

Figure 2. Schematic diagram of the diamond grain coalescence.

It is therefore expected that two neighboring diamond grains with a slight orientation deviation will have tendency to form a single crystal by means of coalescence. In Figure 1, a 2° mis-orientation has been seen to be compensated. In the experiment, the coalescence occurs at a much earlier stage (after 12nm growth). This early coalescence is clearly desirable for the deposition of high quality film and depends on the average distance of the growing islands or the nucleation density. To determine the highest limitation of the mis-orientation compensation, further investigations are necessary so as to claim if the growth of single crystalline diamond films is in principle realizable.

The simulation of the interaction of two diamond grains

To understand the mechanism and find the driving force of coalescence, various simulations of nucleations and growths of two diamond grains on the silicon substrate have been performed. The simulations were modeled with 173 silicon as the substrate and performed by using molecular orbital PM3 method. Each intermediate atomic cluster structure in the step-by-step deposition has been geometrically optimized in the calculations. The atomic morphology of the diamond grains can be observed in real space and at molecular level and the orientation and the bonding tendency between the diamond grains can be monitored.

Figure 3 represents the morphology of two oriented diamond grains nucleated on the (001) silicon surface with the 3-to-2 lattice matching between the diamond and silicon substrate. Pentagon and heptagon units are necessary in the interface for the lattice matching [27,28]. Due to the lattice residual mismatch and the repulsion between the boundary atoms of the two diamond grains, the boundary of the two grains is slightly wider than the space for forming a diamond atomic unit for the coalescence of the two diamond grains. However, it is possible in the experiment that a hydrocarbon species deposits to bridge the edges of the two diamond grains to form weaker C-C bonds with the edge carbon atoms of the two grains. After the structural relaxation, the two diamond grains may be pulled slightly closer. More perfect structure near the coalescence point can be achieved during the further deposition and larger strength for pulling the two diamond grains together will be gained, which may result in lattice strain inside the two diamond grains and the interface. As a result, a larger single diamond grain may be formed with one or both of the starting diamond grains being slightly tilted. The lattice strain may result in multi-twin as shown in Figure 1.

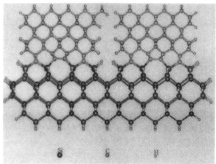

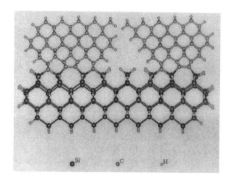

Figure 3. The growth of two diamond grains without tilt.

Figure 4. The growth of two diamond grains with one of them tilted (right side one).

Alternatively, if one of the two diamond grains is grown in an unsymmetrical form, tilting may result [28]. Figure 4 shows an example for such a case, where the right-hand diamond grain is tilted with 3.5°. There is a two-fold effect of the tilting: the two diamond grains are closer and the carbon atoms in the boundary between the two grains are displaced. The later factor is obviously not good for the deposition of a hydrocarbon species to form bonds simultaneously with the edges of the two diamond grains. However, if the tilt angle is small, the two diamond grains can still have a chance to coalesce.

Distribution of electronic density and bonding regeneration

To investigate the possibility and further the maximum diamond grain tilting for diamond grain coalescence, it is necessary to know how the hydrocarbon species interacts with the diamond surface.

The interaction is mainly the efficient overlap of electronic density of the two parts. The interaction distance is a direct measure for the possibility of the coalescence. By studying the distribution of the cross-sectional electronic density in a plane that includes two out-cell carbon atoms of a C_5H_{12} spherical cluster model, it is estimated that the interaction distance is rather short. Within 1.1Å from the carbon atom the electron density drops to about 0.08. And the full feature of the electronic density distribution is directionally dependent – tetrahedral orientation. It shows that for an efficient interaction the distance of the two parties should be properly short and their relative position should be not far from tetrahedral. So, if one of the two grains is of large tilting the atomic morphology between the two diamond grains will obviously not be suitable for forming a regular covalent carbon-carbon bond. The critical tilt angle is at most a few degrees for a possible coalescence. Because the tilting can be slightly corrected during diamond growth [28], a larger tilt angle will require the growth of thicker film for coalescence to occur. Therefore, to obtain thin single crystalline diamond film, the tilt of diamond grains after nucleation should be sufficiently small. It is recommended that the tilt is controlled within $1 \sim 2^\circ$, which can be achieved by using the current technique [17,18].

CONCLUSION

Diamond grains can be coalesced when the tilt is within a few degrees. The coalescence is brought about by the bonding regeneration between two grains during deposition of subsequent layers. For coalescence to occur, the larger the tilt angle within the critical angle is, the thicker the film must be grown.

ACKNOWLEDGMENTS

This work is supported by the Research Grant Council of Hong Kong (project No. 9040195).

REFERENCES

1. B.V. Spitsyn, L.L. Bouilov and B.V. Derjaguin, J. Cryst. Growth 52, 219(1981).
2. M. Yoshikawa, H. Ishida, A. Ishitani, T. Murakami, S. Koizumi and T. Inuzuka, Appl. Phys. Lett. 57, 428(1990).
3. B.R. Stoner, and J.T. Glass, Appl. Phys. Lett. 60, 698(1992); B.R. Stoner, G.H.M. Ma, S.D. Wolter and J.T. Glass, Phys. Rev. B45, 11067(1992).
4. X. Jiang and C.-P. Klages, Diamond Relat. Mater. 2, 1112(1993).
5. X. Jiang, C.-P. Klages, R. Zachai, M. Hartweg and H.-J. Füsser, Appl. Phys. Lett. 62, 3438(1993).
6. X. Jiang, C.-P. Klages, M. Rösler, R. Zachai, M. Hartweg and H.-J. Füsser, Appl. Phys. A57, 483(1993).
7. S.D. Wolter, B.R. Stoner and J.T. Glass, P.J. Ellis, D.S. Buhaenko, C.E. Jenkins and P. Southworth, Appl. Phys. Lett. 62, 1215(1993).
8. Q.J. Chen, J. Yang and Z.D. Lin, Appl. Phys. Lett. 67, 1853(1995); Q.J. Chen, Y. Chen, J. Yang and Z.D. Lin, Thin Solid Films 274, 160(1996).
9. S.G. Song, C.L. Chen, T.E. Mitchell, L.B. Hackenberger and R. Messier, J. Appl. Phys. 79, 1813(1996).
10. M.G. Jubber and D.K. Milne, Phys. Stat. Sol. (a) 154, 185(1996).
11. M. Schreck and B. Stritzker, Phys. Stat. Sol. (a) 154, 197(1996).

12. Y. Von Kaenel, J. Stiegler, E. Blank, O. Chauvet, Ch. Hellwig and K. Plamann, Phys. Stat. Sol. (a) **154**, 219(1996).
13. C.J. Chen, L. Chang, T.S. Lin and F.R. Chen, J. Mater. Res. **11**, 1002(1996).
14. J.C. Angus and C.C. Hayman, Science **241**, 913(1988).
15. A.T. Collins, Semicond. Sci. Technol. **4**, 605(1989).
16. W.A. Yarbrough and R. Messier, Science **247**, 688(1990).
17. X. Jiang and C.L. Jia, Appl. Phys. Letters **69**, 3902(1996).
18. X. Jiang and C.L. Jia, J. Appl. Phys. **83** (5), 2511 (1998).
19. X. Jiang and C.L. Jia, Appl. Phys. Lett. **67**, 1197(1995).
20. C.L. Jia, K. Urban and X. Jiang, Phys. Rev. B**52**, 5164(1995).
21. Q.J. Chen, L.X. Wang, Z. Zhang, J. Yang, Z.D. Lin, Appl. Phys. Lett. **68**, 176 (1996).
22. D.A. Tucker, D.-K. Seo, M.-H. Whangbo, F.R. Sivazlian, B.R. Stoner, S.P. Bozeman, A.T. Sowers, R.J. Nemanich and J.T. Glass, Surf. Sci. **334**, 179(1995).
23. X. Jiang and C.-P. Klages, Phys. Status Solidi A**154**, 175(1996).
24. J.J.P. Stewart, J. Comput. Chem. **2**, 209(1989).
25. M.J.S. Dewar and W.J. Thiel, J. Am. Chem. Soc. **99**, 4899(1977).
26. R.Q. Zhang, W.L. Wang, J. Esteve and E. Bertran, Appl. Phys. Lett **69**, 1086(1996).
27. R.Q. Zhang, W.L. Wang, J. Esteve and E. Bertran, Thin Solid Film, in press; and their following work.
28. X. Jiang, R.Q. Zhang, G. Yu and S.T. Lee, unpubmitted.

THE ORIGIN OF MIS-ORIENTED DIAMOND GRAINS NUCLEATED DIRECTLY ON (001) SILICON SURFACE

R.Q. Zhang*, W.J. Zhang*, C. Sun*, X. Jiang** and S.-T. Lee*
*Department of Physics and Materials Science, City University of Hong Kong, Hong Kong, China
** Fraunhofer-Institut fuer Schicht- und Oberflaechentechnik (FhG-IST), Bienroder Weg 54 E, D-38108 Braunschweig, Germany

ABSTRACT

The origin of mis-oriented diamond grains frequently observed in heteroepitaxial diamond films on (001) silicon surfaces was studied. By statistically analyzing the in-plane rotation angles of diamond grains in scanning electron microscopy observations, it was found that the distribution of the grain orientation is not random and two satellite distribution peaks at about 20° and 30° accompany the main distribution peak at zero degree referenced to the <110> direction of substrate. The interface structure corresponding to the main distribution peak at zero degree of oriented diamond growth has been proposed in our previous studies. In this study, our molecular orbital PM3 simulation of a step-by-step diamond nucleation further reveals two other metastable diamond/silicon interfacial structures. The orientations of the corresponding diamond grains are parallel to the (001) silicon surface but with in-plane rotations of 20° and 30° respectively with respect to the <110> direction. We relate these two mis-oriented growths to the two satellite peaks of grain orientation distribution. Based on this study, the possibility in experiment to reduce the formation of mis-oriented configurations and to obtain a perfectly oriented diamond growth is discussed.

INTRODUCTION

Oriented heteroepitaxial diamond film can be synthesized nowadays on inexpensive silicon (001) and (111) substrates by means of chemical vapor deposition (CVD) techniques [1-9]. Although the films are normally polycrystalline, with enhanced coalescence and overgrowth [10,11] for a film with very small angle misorientation or tilting, the desired single crystal diamond film is hopefully obtained. At present, the reduction of the misoriented diamond grains is one of the crucial factors to reach such a goal so as to fulfil the promising application of the diamond film for electronics purposes [12-14]. An exploration of the origin of the misoriented grains becomes a centrally important task.

Beside the oriented grains that are the majority in the heteroepitaxial diamond film, there are usually some misoriented ones in the observation. Normally, one may think that these grains are randomly misoriented just as in the case of the growth with nucleation on an amorphous base or interlayer. For such a case, the obvious way to reduce the misoriented configuration is to control the experimental parameters carefully so that the etching effect of the hydrogen ion in the plasma on non-diamond phases is optimized while maintaining the low etching rate of the diamond grains. However, whether or not the misoriented diamond grains are randomly distributed requires demonstration so as to reveal the actual mechanism of the misorientation.

In this study, we have statistically analyzed the rotation angles of the CVD diamond grains in scanning electron microscopy (SEM) observations, and have found that the distribution of the grain in-plane orientation is not random and that two satellite distribution peaks at about 20° and 30° accompany the main distribution peak at zero degree with respect to the substrate <110> direction. This strongly indicates that the growth of these misoriented diamond grains in the range of the distribution peaks must relate to nucleation directly on the silicon substrate or via a SiC interlayer, since the growth on an amorphous substrate or domain should assume a random orientation. We also note that during growth the surrounding of such a misoriented grain may influence the orientation. However, such influence would not result in large modification of the grain orientation due to the growth being mainly perpendicular to the substrate or overgrown in some condition as has been observed experimentally. As a result, the orientation of the misoriented grain may be mainly around the orientation of a certain configuration with possibly slight deviation due to the influence by the surrounding or the lattice strain near the diamond grain. On the other hand, if growth occurs via a SiC interlayer the starting configurations for the nucleation with rotations are still similar to the cases for nucleation directly on the substrate.

Our study hence concentrates on diamond nucleation and growth directly on the silicon (001) surface. In this study, molecular orbital PM3 simulation of a step-by-step diamond nucleation is used to further reveal these other metastable structural configurations of diamond nucleation and growth. Therefore, the possibility in experiment to reduce the formation of misoriented configurations can be explored so as to provide information for obtaining a perfectly oriented diamond growth.

EXPERIMENTS AND THEORETICAL APPROACH

Our diamond film was deposited by microwave plasma CVD using the well-known two-step process [15]. The film thickness is about $5\mu m$. In the first step of the deposition, heterogeneous nucleation of [001] oriented diamond crystallites was achieved *in situ* on a 2-in n-type (001) silicon wafer by applying a negative bias potential to the substrate. The second step was an established diamond growth process. The morphology of the film was analyzed by SEM.

For the theoretical work, the PM3 parametrization [16] of the MNDO semi-empirical hamiltonian [17] was used in this study for all calculations. *Ab initio* or other higher level calculations were not used due to the limitation of computation for the presently treated large system. PM3 has been successfully applied to the studies of the diamond growth processes [18,19], structural properties of hydrogenated silicon-carbon alloys [20], chemisorbed state of benzene on Si(100)-(2x1) surface [21], as well as to many other systems that contain elements in a wide range. These calculations have interpreted or predicted interesting experimental phenomena. In most cases, PM3 was found superior to other semi-empirical methods due to the use of a much larger database of compounds in the parametrization, as well as a better fitting algorithm for the parameters [16]. The geometrical parameters produced by PM3 for graphite, diamond [19] and hydrogenated silicon-carbon alloys [20] are very close to the experimental values. In fact, such a theoretical approach has been used in our previous studies of various topics for the diamond nucleation and growth on silicon substrates [22-24]. In this work, all of the cluster models have been geometrically optimized, i.e. energetically relaxed.

RESULTS AND DISCUSSIONS

Figure 1 shows a representative image of the morphology of our diamond film. Most of the grains in this film show their in-plane orientations almost parallel to the <110> with possibly slight orientation deviations. However, there are still many grains misoriented. Among the many thousands of diamond grains that have been considered, 661 of them have been found with orientation deviations larger than ten degrees with respect to the <110> direction.

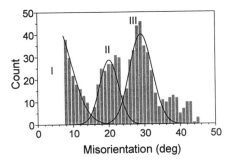

Figure 1. A representative SEM image of morphology of the CVD diamond film.

Figure 2. Distribution of grain orientations of the CVD diamond film.

Figure 2 gives the statistically analyzed result, where the bar chart shows the counted numbers of diamond grains at different angles while the solid curves give the fitting results to Gauss distributions. The orientation distribution of the diamond grain is classified into three ranges: I, II and III. Range I was estimated and represents the oriented diamond grains which actually have not been carefully studied in this work but are familiar to everybody as much work has given such a feature. Peak II and peak III must relate to the nucleation of diamond grain directly on the silicon substrate, since the fine distribution features are clear, fitted well to Gauss distributions and far from random. Actually, during our statistical analysis, almost every image that we selected gives similar peak feature. The feature in Figure 2 is actually the sum for these different images. We note that the error in our measurement of the grain orientation is within $1°$. In the angular range larger than peak III, there is no obvious characteristic peak, indicating no metastable configuration in the related nucleation. Thus, the $45°$ rotated configuration proposed in the early work of Verwoerd [25] but dismissed in his later work can not be observed in our diamond film either.

The diamond nucleation and growth for an oriented film have been theoretically studied previously at atomic level by one of the present authors [22,23]. Similarly, the mechanism of grain tilting for heteroepitaxial oriented diamond film has been theoretically studied recently [24]. These studies have mainly considered the step-by-step deposition procedure of carbon species directly on silicon substrate. The structural configurations of diamond nucleation and the interface were revealed by determining the adsorption sites of hydrocarbon species on silicon surfaces by means of the PM3 calculation [22]. At the initial stage of diamond nucleation, an adsorbed CH_2 may be bonded with and bridging two surface Si atoms, while two hydrocarbon species or a single C-C species may form a heptagon with the neighboring surface Si atoms. These two kinds of micro-configurations provide basis for diamond nucleations in our simulations, some of which have been reported previously as above-mentioned.

In this study, a cluster model composed of 109 silicon atoms with hydrogen saturated boundary was selected to simulate the silicon (001) surface. Figure 3 gives the top view of such a substrate with the initial deposition of two neighboring C-C configurations mentioned above. However, these two C-C configurations are slightly different from those in our previous studies: Due to the presence of a metastable state of such configuration, a perturbation may result in the rotated local structure as shown in the Figure 3. So, the C-C configuration is no longer parallel to the [110] direction.

Now, let's consider the diamond nucleation based on such a configuration. If a hydrocarbon species bridges the two nearest carbon atoms belonging to the different C-C configurations by abstracting or desorbing the saturation hydrogen atoms, a nucleation point for a (001) growth is provided. The continuous deposition on such a nucleation point may result in the diamond growth as shown in Figure 4, where the rotation of the grown diamond grain is obvious. For this case, the in-plane rotation angle is about 20°. We have made various similar simulations with different size of substrate or different size of diamond grain, and the rotation angles have been found quite similar.

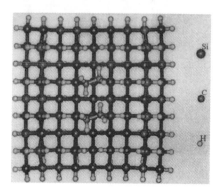 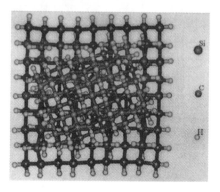

Figure 3. The deposited C-C configurations at the first carbon layer on the Si(001) surface with slight rotation.

Figure 4. Top view of the diamond grain 20° misoriented due to the nucleation on rotated C-C configurations as in figure 3.

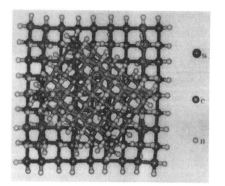 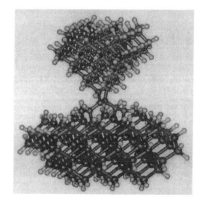

Figure 5. (a) Top view (left figure) and (b) side view (right figure) of the diamond grain 30° misoriented due to the nucleation on rotated C-C configurations as shown in figure 3.

However, if a C-C species or two hydrocarbon species bridge such two carbon atoms belonging to the different C-C configuration of the first deposition layer by abstracting or desorbing the saturated hydrogen atoms, an alternative nucleation point is formed. The diamond grain grown on such a nucleation point is illustrated in Figure 5 (a) and (b), where a top view is to show the rotation of the diamond grain and the side view in figure 5(b) is to give the interface structure and shows the bonding feature between the diamond grain and the silicon substrate. The in-plane rotation of this kind of diamond grain is about 30° with respect to the <110> direction.

By comparing the in-plane rotation angles of the diamond grains grown on these two kinds of nucleation points, it should not be surprising if we relate such two configurations to the two-peak feature in the orientation distribution of the diamond grain observed in our SEM for the CVD diamond film. Although there are many possible nucleation configurations on the substrate, most of the nuclei with poor bonding can be etched out by hydrogen radical/ion. So, the configurations of nuclei which are suitable for diamond growth are limited. The diamond grains grown according to the nucleation mechanism in the above simulations are obvious with quite strong bonding so that such growth should be present in the experiment.

Energetically, the formation of the cluster model with diamond in-plane rotation of 20° is about 0.6eV higher than that of 30° rotation in our calculations. It is possibly the reason why the distribution peak around 30° is of a greater height and a larger area, which means that the formation of the later configuration is more favorable during nucleation. This agreement can also be regarded as a supportive evidence of our models of diamond mis-orientation.

On the other hand, we would not rule out other possibilities which may contribute to the two orientation distribution peaks and will make further exploration for revealing the complete mechanism of diamond grain misorientations. With the present understanding of the diamond nucleation at the molecular level, we may have certain confidence to think about how to reduce the formation of the misoriented diamond grains. First, the substrate should be made sufficiently smooth so that a good basis for nucleation is provided. Secondly, the nucleation condition should be steady so that perturbation which may result in the rotation of the C-C configuration in the first deposition layer can be reduced. A new CVD method involving nucleation by low-pressure electron cyclotron resonance plasma that is being developed in our laboratory and has already produced supportive evidence for hetero-epitaxy [26] should be suitable for such a purpose.

CONCLUSION

The orientation of the misoriented diamond grains in our CVD heteroepitaial diamond films does not distribute randomly. Two distribution peaks of the in-plane orientation angle locate respectively at about 20° and 30° with respect to the <110> direction of the substrate. The grains corresponding to these peak areas must nucleate directly on the silicon substrate and can be related to two nucleation configurations revealed in our theoretical simulations. Diamond growth based on these two nucleation configurations gives diamond grains being rotated about 20° and 30° respectively. The good agreement between the experimental observation and the theoretical finding indicates that our theoretical approach is quite successful.

ACKNOWLEDGMENTS

This work is supported by the Research Grant Council of Hong Kong (project No. 9040195).

REFERENCES

1. X. Jiang and C.-P. Klages, Diamond Relat. Mater. **2**, 1112(1993).
2. X. Jiang, C.-P. Klages, R. Zachai, M. Hartweg and H.-J. Füsser, Appl. Phys. Lett. **62**, 3438(1993).
3. X. Jiang, C.-P. Klages, M. Rösler, R. Zachai, M. Hartweg and H.-J. Füsser, Appl. Phys. **A57**, 483(1993).
4. S.D. Wolter, B.R. Stoner and J.T. Glass, P.J. Ellis, D.S. Buhaenko, C.E. Jenkins and P. Southworth, Appl. Phys. Lett. **62**, 1215(1993).
5. Q.J. Chen, J. Yang and Z.D. Lin, Appl. Phys. Lett. **67**, 1853(1995); Q.J. Chen, Y. Chen, J. Yang and Z.D. Lin, Thin Solid Films **274**, 160(1996).
6. S.G. Song, C.L. Chen, T.E. Mitchell, L.B. Hackenberger and R. Messier, J. Appl. Phys. **79**, 1813(1996).
7. M.G. Jubber and D.K. Milne, Phys. Stat. Sol. (a) **154**, 185(1996).
8. M. Schreck and B. Stritzker, Phys. Stat. Sol. (a) **154**, 197(1996).
9. Y. Von Kaenel, J. Stiegler, E. Blank, O. Chauvet, Ch. Hellwig and K. Plamann, Phys. Stat. Sol. (a) **154**, 219(1996).
10. X. Jiang and C.L. Jia, Appl. Phys. Lett. **69**, 3902 (1996).
11. X. Jiang and C.L. Jia, J. Appl. Phys. 83 (5), 2511 (1998).
12. J.C. Angus and C.C. Hayman, Science **241**, 913(1988).
13. A.T. Collins, Semicond. Sci. Technol. **4**, 605(1989).
14. W.A. Yarbrough and R. Messier, Science **247**, 688(1990).
15. X. Jiang and C.-P. Klages, Phys. Status Solidi **A154**, 175(1996).
16. J.J.P. Stewart, J. Comput. Chem. **2**, 209(1989).
17. M.J.S. Dewar and W.J. Thiel, J. Am. Chem. Soc. **99**, 4899(1977).
18. B.H. Besler, W.L. Hase and K.C. Hass, J. Phys. Chem. **96**, 9369(1992).
19. P. Deák, A. Gali, G. Sczigel and H. Ehrhardt, Diamond Rel. Mater. **4**, 706(1995).
20. R. Colle and K.K. Stavrev, J. Solid State Chem. **117**, 427(1995).
21. H.D. Jeong, S. Ryu, Y.S. Lee and S. Kim, Surf. Sci. **334**, L1226(1995).
22. R.Q. Zhang, W.L. Wang, J. Esteve and E. Bertran, Thin Solid Film, in press; and their following work.
23. R.Q. Zhang, W.L. Wang, J. Esteve and E. Bertran, Appl. Phys. Lett. **69**, 1086(1996).
24. X. Jiang, R.Q. Zhang, G. Yu and S.T. Lee, Phys. Rev. B, submitted.
25. W.S. Verwoerd, Surf. Sci. **304**, 24 (1994); and his related work.
26. C. Sun, W.J. Zhang, C.S. Lee, I. Bello, and S.T. Lee, Diamond Rel. Mater., submitted.

MODELING OF SUBSTRATE BIAS EFFECT ON THE COMPOSITIONAL VARIATIONS IN SPUTTER-DEPOSITED TiB$_{2+x}$ DIFFUSION BARRIER THIN FILMS

M. SINDER, G. SADE AND J. PELLEG
Department of Materials Engineering, Ben-Gurion University of the Negev, Beer-Sheva, 84105, Israel

ABSTRACT

   Sputter-deposited titanium boride diffusion barrier layers have been found to be boron enriched when r.f. substrate bias was applied. In the present experiments titanium boride was deposited by co-sputtering from Ti and B pure targets in Ar discharge and the voltage of r.f. self-bias was in the range of 100 - 250 V. Films deposited were found by Auger electron spectroscopy to be B enriched with increasing bias voltage at constant Ti and B sputtering rates. A model of the sputter-deposition conditions was developed to predict the composition and the thickness of the growing film. The model explains the experimental results indicating that B enrichment is mainly a result of differential resputtering of the components from the growing film by energetic Ar ions captured from the r.f. discharge.

INTRODUCTION

   Because of their unique physical and technological properties, thin films of titanium boride are one of the most attractive materials for diffusion barriers in very large scale integrated devices. Although this possibility was suggested as far as 1969 [1], not so many works have been reported since then on using TiB$_2$ for this application. Nicolet [2] has indeed suggested TiB$_2$ as a possible diffusion barrier in silicon technology. Blom et al. [3], Shappirio et al. [4] examined the use of TiB$_2$ thin film as diffusion barrier against Al, and Choi et al. [5] examined its applicability against Cu diffusion. Thin films of TiB$_2$ have been fabricated by direct evaporation [6], chemical vapor deposition [7, 8, 9, 10, 11], reaction of boron-metal thin film couples [12, 13], reactive sputtering [14, 15], sputtering from TiB$_2$ target [4,16,17], co-sputtering from TiB$_2$ and B targets [18, 19, 20] and, for maximum flexibility, by co-sputtering from elemental Ti and B targets [21]. The choice of co-sputtering from elemental targets was found preferable, since the ratio of the components can be varied according to the research objective. When sputtering technique is in use for film deposition, to achieve adequate step coverage of highly packed, multilevel interconnections in VLSI, negative bias has to be applied on a substrate during sputtering [22]. Substrate bias causes resputtering and redeposition of deposited atoms by ion bombardment, as well as enhances the adatom mobility, which can result in crystallization. Indeed, Shikama et al. [19, 20] observed as-deposited crystalline TiB$_2$, when crystallization was stimulated by negative bias during the deposition and a significant enhancement of step coverage has been demonstrated using substrate bias, either d.c. or r.f. [23, 24]. Besides, deposition with bias improves film purity by an "atom peening" process, although, for any given deposition time, bias sputtering results in thinner deposited films [25]. Low energy ion bombardment of thin films during growth occurring when bias is applied, was the subject of many studies. Changes in nucleation characteristics [26] as well as changes in the morphology [27] and composition [28, 29] of the films have been reported. In our previous study [30] improved diffusion barrier properties were found in amorphous TiB$_2$ deposited by co-sputtering from elemental targets with r.f. bias. This effect was associated with the enrichment with B of the TiB$_2$ film [31]. However, the effect of substrate bias on the compositional variation still remains unclear. The objective of this work is to estimate the relationship between substrate bias and composition of the films formed by co-sputtering from elemental targets.

EXPERIMENT

Films were deposited in a "Cook" magnetron sputtering system by co-sputtering using elemental Ti (99.99%) and boron (99.9%) targets, which were sputter etched before depositions. The base pressure before depositions was typically ~10 μPa. Sputtering was carried out in 0.5 Pa of high purity Ar (99.999%). Typical input power to the Ti and B targets were 320 W r.f. and 60 W dc respectively, whereas r.f. bias voltages varied from –100 to –250 V. The final film thickness was about 30 nm. No heating was applied to the substrate during the depositions.

The composition of the as-deposited films was determined from Auger depth profiling. The measured intensities of the 418 eV (LMM) titanium peak, the 179 eV (KLL) boron peak, the 92 eV (LMM) silicon peak, the 272 eV (KLL) carbon peak, and the 512 eV (KLL) oxygen peak from data accumulated in the derivative mode were used to compute the atomic concentration. Sensitivity factors of the metallic elements (Ti and B) given by Davis et al. [32] were used to quantify the B/Ti ratio, since neither TiB_2 nor boron standards were available for the calibration. A 2 keV, 0.5 μA argon ion beam was used to etch by sputtering of the sample surface. The gas pressure for this operation was 2×10^{-3} Pa, whereas the base pressure of the system was below 5×10^{-8} Pa. The electrical resistivity of the films was calculated from the sheet resistance obtained as routine measurements by a four-point probe and the film thickness was measured from the cross-sectional transmission electron microscopy.

RESULTS AND DISCUSSIONS

The variation in the composition with substrate bias of co-sputtered titanium boride is shown in Fig. 1.

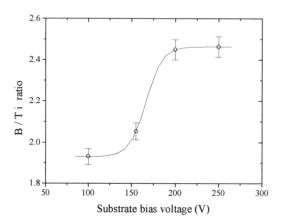

Fig. 1. Chemical composition (B/Ti ratio) of titanium boride films as a function of r.f. bias voltage.

As can be seen from the graph, the boron content in the deposited films increases with increasing substrate bias voltage. This phenomenon can be explained by preferential resputtering of titanium from the growing film by energetic Ar ions being generated in r.f. discharge induced near the substrate when bias is applied. When a negative bias is applied to the substrates the number of ionized molecules and atoms in the plasma increases. The energetic electrons from the target strike the substrate and cause emission of secondary electrons, which are accelerated into the plasma by the bias field. Since the ionization probability due to electron impact has a maximum at about 100-150 eV for most gases, the electrons accelerated by the bias field are effective in increasing the ionization of the plasma. The total number of Ar ions on the substrate increases. As the bias increases, the energy of the species impinging on the substrate also increases as does the resputtering [29].

A simplified mathematical model has been developed to explain the compositional variations in titanium boride thin films deposited with r.f. bias.

MATHEMATICAL MODEL FOR DEPOSITION WITH R.F. BIAS

Description and assumptions of the model

Deposition of TiB$_2$ by co-sputtering is performed from two independent sources (targets) and we designate the flows of boron and titanium atoms from the targets, which land on the substrate by J_B^+ and J_{Ti}^+, respectively. Application of r.f. bias creates a flow of energetic Ar ions which are accelerated towards the substrate, resulting in resputtering of the growing layer. Let us designate by J_{Ar} the flow of Ar ions. The term "flow" means the number of atoms condensed per unit time per unit surface. We also designate the film growth rate of TiB$_2$ by V, and the number of B and Ti atoms in unit volume of TiB$_2$ film by ρ_B and ρ_{Ti}, respectively. J_B^- and J_{Ti}^- are the flows of boron and titanium atoms respectively which are resputtered by energetic Ar ions from the growing TiB$_2$ film. Suppose that the state of simultaneous sputtering and resputtering is stationary, (i.e., it is not a time dependant process). Then, from the atom flow balance on the surface of the growing layer we obtain:

$$\rho_B V = J_B^+ - J_B^- \tag{1}$$

$$\rho_{Ti} V = J_{Ti}^+ - J_{Ti}^- \tag{2}$$

Assuming that the atoms in the TiB$_2$ film are densely packed we can write:

$$\rho_B \Omega_B + \rho_{Ti} \Omega_{Ti} = 1 \tag{3}$$

where Ω_B, and Ω_{Ti} are the volumes per atom of B and Ti in the TiB$_2$ layer.

From the theory of sputtering [33, 34], the resputtering rate (in our case the flows J_B^- and J_{Ti}^-) is proportional to the flow of the incident Ar ions and the number of B and Ti atoms on the sputtered surface. Thus we have:

$$J_B^- = S_B \rho_B J_{Ar} \tag{4}$$

$$J_{Ti}^- = S_{Ti} \rho_{Ti} J_{Ar} \tag{5}$$

where S_B and S_{Ti} are the sputtering yields of boron and titanium respectively. Thus, equations (1) – (5) are the model equations. From these equations and knowing the values $J_B^+, J_{Ti}^+, \Omega_B, \Omega_{Ti}, S_B$

and S_{Ti} we can determine the growth rate V and the composition of the deposited titanium boride film as a function of the flow of incident ions (J_{Ar}).

Analysis of the model

Multiplying equations (1) and (2) by Ω_B and Ω_{Ti} respectively, adding them and using equations (3)-(5) we have:

$$\frac{J_B^+\Omega_B}{V+J_{Ar}S_B}+\frac{J_{Ti}^+\Omega_{Ti}}{V+J_{Ar}S_{Ti}}=1 \tag{6}$$

The schematic graph of equation (6) (V versus J_{Ar}) is shown in Fig. 2. This graph shows a monotonic decrease of the film growth rate, V, with increasing flow of the incident ions, J_{Ar}

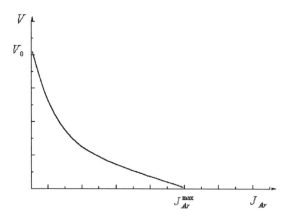

Fig. 2. Schematic graph of TiB$_2$ deposition rate as a function of flow of Ar incident ions.

Analyzing equation (6) we can make some conclusions:
1. In the absence of flow of Ar incident ions, $J_{Ar}=0$ (no bias is applied to the substrate), the film growth rate is maximal: $V_0=J_B^+\Omega_B+J_{Ti}^+\Omega_{Ti}$.
2. The flow of the energetic Ar ions when no deposition occurred, i.e., the deposition rate equals the resputtering rate ($V=0$ in our case):

$$J_{Ar}^{max}=\frac{J_B^+\Omega_B}{S_B}+\frac{J_{Ti}^+\Omega_{Ti}}{S_{Ti}} \tag{7}$$

From equations (1), (2) and using equations (4), (5) we can find the composition of the deposited titanium boride as a function of the flow of the Ar incident ions

$$\frac{\rho_B}{\rho_{Ti}}=\left(\frac{J_B^+}{J_{Ti}^+}\right)\cdot\frac{V+S_{Ti}J_{Ar}}{V+S_B J_{Ar}} \tag{8}$$

where V as a function of J_{Ar} is determined from equation (6).

The composition of the deposited titanium boride film with bias vs. J_{Ar}, is a monotonically increasing curve as shown in Fig. 3.

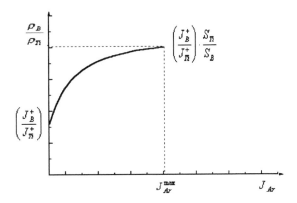

Fig. 3. Schematic graph of TiB$_2$ composition as a function of flow of Ar incident ions.

The compositional variation of titanium boride as a function of Ar incident ion flow, when the deposition rate is large enough, i.e., the influence of the flow of Ar ions is not significant on the TiB$_2$ deposition rate, can be determined approximately from (6), (8):

$$\left(\frac{\rho_B}{\rho_{Ti}}\right) - \left(\frac{\rho_B}{\rho_{Ti}}\right)_{J_{Ar}=0} = \left(\frac{\rho_B}{\rho_{Ti}}\right)_{J_{Ar}=0} \cdot \frac{(S_{Ti}-S_B)}{V_0} \cdot J_{Ar} \qquad (9)$$

Thus, if $S_{Ti} > S_B$, i.e., sputtering yield of titanium exceeds the sputtering yield of boron, the deposited titanium boride layer will be enriched by the element with the smaller sputtering yield, boron in our case.

As can be seen from (6) increasing J_{Ar} reduces the deposition rate of TiB$_2$ film (V) and consequently from (8) ρ_B/ρ_{Ti} is going to its maximum value, as shown on Fig. 3.

$$\left(\frac{\rho_B}{\rho_{Ti}}\right)_{max} = \frac{J_B^+}{J_{Ti}^+} \cdot \frac{S_{Ti}}{S_B} = \left(\frac{\rho_B}{\rho_{Ti}}\right)_{J_{Ar}=0} \frac{S_{Ti}}{S_B} \qquad (10)$$

CONCLUSION

It was shown in this work that a negative substrate bias affects the composition of Ti-B deposits. An increase in substrate bias increases the boron content in the film. The mathematical model developed explains to a first approximation the experimentally obtained composition variations in the titanium boride film resulting from the application of an r.f. bias. In order to maintain a desired B/Ti ratio, the applied bias has to be chosen as indicated in Fig.3. The model also predicts that in order to produce high quality TiB$_2$ films at the desired B/Ti ratio a moderate bias voltage should be applied to reduce its influence on the compositional variations, recalling that high bias voltage results in reduction of the deposition rate.

REFERENCES

1. C.W. Nelson, in 1969 Hybrid Microelectronics Symposium (International Society for Hybrid Microelectronics, Hicks Printing Co., Dallas, TX 1969).
2. M.-A. Nicolet, Thin Solid Films **52,** 415 (1978).
3. H.-O. Blom, T. Larsson, S. Berg and M. Ostling, J. Vac. Sci. Technol. **A7**(2) 162 (1989).
4. J. R. Shappirio, J. J. Finnegan, R. A. Lux, J. Vac. Sci. Technol. **B4**(6) 1409 (1986).
5. C. S. Choi, G. A. Ruggles, A. S. Shah, G. C. Xing, C. M. Osburn, and J. D. Hunn, J. Electrochem. Soc. **138**(10) 3062 (1991).
6. A.E. Kaloyeros, M. P. Hoffman, W.S. Williams, A.E. Greene, and J.A. McMillan, Phys. Rev. B **38**(11) 7333 (1988).
7. H.O. Pierson and A.W. Mullendore, Thin Solid Films **72** 511 (1980).
8. L. M. Williams, Appl. Phys. Lett. **46** 43 (1985).
9. F.A. Otter, G.B. Amisola, W.C. Roman, and S.O. Hay, J. Vac. Sci. Technol.A **10**(4) 2796 (1992).
10. C.S. Choi, G.C. Xing, G.A. Ruggles, and C.M. Osburn, J. Appl. Phys. **69** 7853 (1991).
11. C.S. Choi, G.A. Ruggles, C.M. Osburn, and G.C. Xing, J. Electrochem. Soc. **138**(10) 3053 (1991).
12. C. Feldman, F. G. Satkiewicz, and G. Jones, J. Less-Common Met. **79** 221 (1981).
13. C. Feldman, F. G. Satkiewicz, and N. A. Blum, J. Less-Common Met. **82** 183 (1981).
14. T. Larsson, H.-O. Blom, S. Berg, and M. Ostling, J.Vac. Sci. Technol. A **7**(2) 162 (1989).
15. H.-O. Blom, T. Larsson, S. Berg, and M. Ostling, Thin Solid Films **172** 133 (1989).
16. K. R. Pdmanabhan and G. Sorensen, Thin Solid Films **81** 13 (1981).
17. J. Chen, J.A. Barnard, Materials Science and Engineering A **191** 233 (1995).
18. J. G. Ryan, S. Roberts,G. J. Slusser, and E. D. Adams, Thin Solid Films **153** 329 (1987).
19. T. Shikama, Y. Sakai, M. Fukutomi, and M. Okada, Thin Solid Films **156** 287 (1988).
20. T. Shikama, Y. Sakai, M. Fujitsuka, Y. Yamauchi, H. Shindo, and M. Okada, Thin Solid Films **164** 95 (1988).
21. E.Matsubara, Y.Waseda, S.Takeda and Y.Taga, Thin Solid Films, **186** L33 (1990).
22. F. Smith, F.T. Zold and W. Class, Thin Solid Films, **96** 291 (1982).
23. J.F. Smith, Solid State Technol., **27** 135 (1984).
24. D.W. Skelly and L.A. Gruenke, J. Vac. Sci. Technol., **4** 457 (1986).
25. J.R. Shappirio, J.J. Finnegan, R.A. Lux, and D.C. Fox, Thin Solid Films, **119** 23 (1984).
26. M. Marinov, Thin Solid Films, **46** 267 (1977).
27. D.M. Mattox and G.J. Kominiak, J.Vac.Sci.Technol., **9** 528 (1972).
28. L.I. Maissel and P.M. Schaible, J. Appl. Phys., **36** 237 (1965).
29. J.-E. Sundgren, B.-O. Johansson, H.T.G. Hentzell, and S.-E. Karlsson, Thin Solid Films, **105** 385 (1983).
30. G. Sade and J. Pelleg, Applied Surface Science **91,** 263 (1995).
31. G. Sade, J. Pelleg, (Mater. Res. Soc. Symp. Proc. **402,** Pittsburgh, PA 1996) 131.
32. L.E. Davis, N.C. MacDonald, P.W. Palmberg, G.E. Riach, and R.E. Weber, Handbook of Auger Electron spectroscopy (Physical Electronics Industries, Eden Prairie, MN, 1976).
33. P. Sigmund, Phys. Rev. **184** 383 (1969).
34. P. Sigmund, Phys. Rev. **187** 768 (1969).

FAILURE MECHANISMS OF PARTICULATE
TWO-PHASE COMPOSITES

T. ANTRETTER*, F. D. FISCHER*
* Institute of Mechanics, Montanuniversität Leoben, Franz-Josef-Straße 18, A-8700 Leoben, Austria, e-mail: antrette@grz08u.unileoben.ac.at

ABSTRACT

In many composites consisting of hard and brittle inclusions embedded in a ductile matrix failure can be attributed to particle cleavage followed by ductile crack growth in the matrix. Both mechanisms are significantly sensitive towards the presence of residual stresses.

On the one hand particle failure depends on the stress distribution inside the inclusion, which, in turn, is a function of various geometrical parameters such as the aspect ratio and the position relative to adjacent particles as well as the external load. On the other hand it has been observed that the absolute size of each particle plays a role as well and will, therefore, be taken into account in this work by means of the Weibull theory. Unit cells containing a number of quasi-randomly oriented elliptical inclusions serve as the basis for the finite element calculations. The numerical results are then correlated to the geometrical parameters defining the inclusions. The probability of fracture has been evaluated for a large number of inclusions and plotted versus the particle size. The parameters of the fitting curves to the resulting data points depend on the choice of the Weibull parameters.

A crack tip opening angle criterion (CTOA) is used to describe crack growth in the matrix emanating from a broken particle. It turns out that the crack resistance of the matrix largely depends on the distance from an adjacent particle. Residual stresses due to quenching of the material tend to reduce the risk of particle cleavage but promote crack propagation in the matrix.

INTRODUCTION

The particles in a composite typically add stiffness to the material whereas the matrix ensures a sufficiently high overall toughness. Depending on the material properties of the two phases damage initiation can be attributed to various mechanisms. In many instances it has been found that cracks originate from broken inclusions and propagate into the matrix, thus causing other inclusions to break as well. Again, depending on the characteristics of the interface, the crack may find a path around the inclusions rather than through them. This study confines itself to the investigation of materials with a perfect interface, i.e. debonding will not be taken into account. Furthermore, it is assumed that the inclusions fail prior to the surrounding matrix. This assumption is not unreasonable, given the fact that stiff inclusions exhibit high stresses over a larger area than the surrounding soft matrix.

In many practically used two phase composites it has been observed that large particles tend to be more susceptible to cleavage than small ones. This phenomenon cannot be explained by means of a principal stress criterion only since the distribution of σ_1 does not depend on the length scale of the problem (as long as classical local continuum mechanics is applied). A concept providing a tool for handling size effects was developed by Weibull ([1]): For inhomogeneous, uniaxial stress states $\sigma(\underline{x})$, with $\underline{x}$ denoting the coordinates of a material point, the probability of failure is

Mat. Res. Soc. Symp. Proc. Vol. 529 © 1998 Materials Research Society

$$P_f(V) = 1 - exp\left[-\frac{1}{V_0} \int_V \left(\frac{\sigma(\underline{x})}{\sigma_0}\right)^m dV\right], \qquad (1)$$

where V_0 is an arbitrary scaling parameter, σ_0 the characteristic strength of the material and m a material parameter called the Weibull modulus which is basically determined by the scatter of the measured strength data in a series of fracture tests.

The integration is performed over regions exposed to tensile stresses only. Compressive stresses are not damaging. As far as the particles in a composite are concerned, one has to keep in mind that an externally applied tension σ_∞ causes a rather irregular stress distribution in each inclusion represented by the stress tensor $\underline{\sigma}(\underline{x})$. Since a scalar expression is required for σ in eqn. (1) the major principal stress $\sigma_1(\underline{x})$ obtained from $\underline{\sigma}(\underline{x})$, which typically triggers cracking in brittle materials, will be used.

In order to avoid catastrophic failure of the entire composite it is generally desirable for the matrix to absorb part of the energy released during fracture by forming a plastic zone around the crack tip. In that way a lower amount of energy will be available for the creation of new surfaces. In other words, the material inhibits stable crack growth. To what extent crack growth can be inhibited and possibly stopped by the matrix does not only depend on the matrix properties themselves. It is also a function of the arrangement of surrounding inclusions, which has a considerable influence on the stress distribution inside the matrix.

THE PROBABILITY OF INCLUSION FRACTURE IN RANDOM CONFIGURATIONS

The results and relationships worked out in the following have been obtained through a number of two-dimensional finite element analyses of multi-inclusion unit cells (see [2]) containing around 20 to 40 quasi-randomly distributed elliptical inclusions of various sizes, shapes and orientations. The issue to what extent P_f depends on the particle size V is greatly influenced by the choice of m. For a given ratio $\sigma_\infty/\sigma_0 = 0.8$ and a volume fraction of inclusions of $f_V = 15\%$, Figure 1 displays a distribution of data points for three different values of m. The data points for high values of m are generally below those for low values of the Weibull modulus. Remarkably, also the scatter of the data increases with increasing m. In the limiting case $m \to \infty$ the probability of fracture P_f becomes merely a function of the maximum value of the major principal stress inside each inclusion with σ_0 representing a discrete value for the fracture strength of the material. Eqn. (1) would then only give either $P_f = 0$ or $P_f = 1$ which would appear as two discrete horizontal lines in Figure 1. From that point of view the statistical approach that is employed here can be regarded as a generalization of a fracture criterion based on the major principal stress. On the other hand, as m approaches zero, the influence of the stresses vanishes, and a sharp line giving a one to one correspondence between particle volume and P_f remains.

To each set of data points a fitting curve can be found. Such fitting curves have been evaluated for the inclusion fractions $f_V = 15\%$ and $f_V = 30\%$, respectively, and two different load cases, i.e. purely mechanical load as opposed to mechanical plus superimposed thermal load as shown in Figure 2. The beneficial influence of residual stresses due to quenching of the material becomes evident.

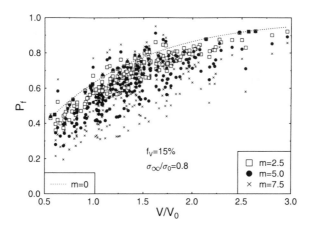

Figure 1: P_f as a function of the particle size V/V_0 for uniaxial tension. $\sigma_\infty/\sigma_0 = 0.8$, $f_V = 15\%$. For $m = 0$ the probability of fracture is merely a function of V/V_0. For $m \to \infty$ the function $P_f(V/V_0)$ degenerates to the two distinct lines $P_f = 0$ and $P_f = 1$.

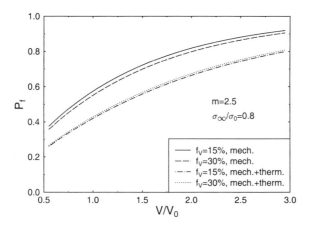

Figure 2: P_f vs. V/V_0 curves for two different volume fractions of particles and two different loading conditions. $\sigma_\infty/\sigma_0 = 0.8$, Weibull modulus $m = 2.5$.

CRACK PROPAGATION IN THE MATRIX

An inclusion is most likely to cleave along a line perpendicular to the direction of the major principal stress, which - for uniaxial tension - essentially matches the direction of the global load ([3]). Similarly, the extension of the crack into the matrix will generally follow that rule even though a brittle fracture criterion, strictly speaking, cannot be applied. If a second inclusion is arranged next to a damaged one as displayed in Figure 3 the crack actually runs into an area of reduced stresses, which leads to an increase in fracture resistance.

In order to facilitate crack initiation in the matrix the model assumes that the first inclusion is already damaged before it is exposed to any external loading. The crack path is determined in the finite element model by a set of nodes along the x-axis that may debond based on a crack tip opening angle (CTOA) criterion ([4,5]). For the given symmetric particle arrangement such a crack path is physically relevant as long as the crack tip has not reached the interface of the second inclusion. Beyond that point the validity of the model is questionable as it does not accommodate any bifurcation phenomena which may occur at the interface. In this paper decohesion of the interface is not dealt with, i.e. once the crack hits the interface it will not proceed any further. Figure 3 shows the model that has been used.

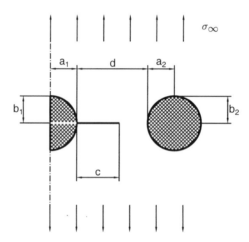

Figure 3: The model consisting of two inclusions embedded in a ductile matrix. Uniaxial tension is applied in y-direction. The axes of the ellipses are aligned with the global coordinates. In an effort to keep the computation time as short as possible the model is assumed to be symmetric with respect to the x- and the y-axis.

The matrix is capable of yielding up to a certain degree. The volume fraction of the inclusions may be as high as 50%, therefore a strong interaction of an existing crack with its surrounding particles can be expected.

The λ vs. crack length curves can be interpreted as crack resistance curves where λ is defined as a load proportionality factor, i.e. the ratio of the applied load σ_∞ to some reference stress (here: σ_0 of the particle). Figure 4 displays the crack resistance curves for four different distances between the inclusions. The slope of each curve indicates the fracture resistance of the particle configuration. It turns out that closely packed inclusions have the highest potential to inhibit crack growth. As soon as the crack hits the interface it cannot be driven ahead by an increase of the external load, so the crack resistance curves degenerate to vertical lines. It should be noted that the calculations were stopped as soon as the probability of fracture of the second inclusion exceeded 99%. For long distances between the inclusions this is the case before the crack ever reaches the interface of inclusion No. 2.

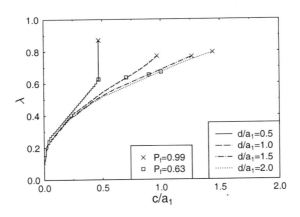

Figure 4: Crack resistance curves for various values of d. $a_1/b_1 = 1, a_2/b_2 = 1$. The two inclusions are equal in size.

Residual stresses after quenching caused by the thermal mismatch of the constituents of the composite ($\alpha_{mat} = 2\alpha_{inc}$ has been assumed here for the coefficients of thermal expansion of the inclusions and the matrix, respectively) generally have a beneficial effect on the inclusions but promote crack growth in the matrix in the case of uniaxial tension. This behavior is illustrated by the crack resistance curves depicted in Figure 5. The slopes of the curves are generally less steep than those for the cases of purely mechanical loading. As much as residual stresses due to thermal strains support the inclusions, they have a detrimental effect on the matrix.

CONCLUSIONS

For the cases of purely mechanical load a higher volume fraction of the particles will both help to reduce the probability of fracture of the carbides and to decrease the overall stress level in the matrix. It has been shown that even though a close arrangement inhibits both crack initiation by particle failure and crack growth into the matrix, it has a lower potential to prevent catastrophic failure of the second inclusion than a wide-spread particle configuration. Despite their beneficial effects on the particles, residual stresses caused by quenching facilitate crack growth in the matrix.

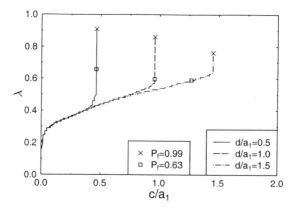

Figure 5: Crack resistance curves for various values of d. Residual stress are assumed: $E\alpha\Delta T/\sigma_0 = 0.63$. $a_1/b_1 = 1$, $a_2/b_2 = 1$. The two inclusions are equal in size.

REFERENCES

[1] Danzer, R.: A General Strength Distribution Function for Brittle Materials. *J. Europ. Ceram. Soc.* **10,** (1992) 461–472.

[2] Antretter, T.: Micromechanical Modeling of High Speed Steel. *Doctoral Thesis*, Montanuniversität Leoben, Austria, 1998, p. 48.

[3] Antretter, T. and Fischer, F. D.: Stress State and Failure of Carbides in Tool Steel – A Micromechanical Study. *Progress in Tool Steels* (Eds H.Berns, H.F.Hinz, I.M.Hucklenbroich), 501–509; Schürmann und Klagges, Bochum, Germany (1996).

[4] Shan, G. X., Kolednik, O., Fischer, F. D.: A Numerical Investigation on the Geometry Dependence of the Crack Growth Resistance in CT Specimens. *Int. J. Fract.* **66,** (1994) 173–187.

[5] Shan, G. X., Kolednik, O., Fischer, F. D.: A Numerical Study on the Crack Growth Behavior of a Low and a High Strength Steel. *Eng. Fract. Mech.* **78,** (1996) 335–346.

COMPUTATIONAL MODELLING OF RING-SHAPED MAGNETIC DOMAINS

A.F. Gal'tsev, V.G. Pokazan'ev, and Y.I. Yalishev
Ural State Academy of Railway Transport, Ekaterinburg, 620034, RUSSIA, afg@nis.usart.ru

ABSTRACT

A theoretical research of static stability of peculiar localized domain structures (LDS) in a thin magnetic film with the perpendicular anisotropy is presented. The model describes a system consisting of cylindrical magnetic domain (CMD) and several concentric ring-shaped domains. Such structures arise under influence of the external low frequency (100-1000 Hz) magnetic field applied perpendicular to the film plane and were observed experimentally in 1992. Non-linear singular integro-differential equation for a magnetization distribution is provided by a minimization condition for the system's complete energy local density. Energy dependencies on geometry parameters are calculated numerically. The conditions of magnetostatic stabilization of the simplest CMD-ring system, as well as some of its dynamical properties, are discussed in detail on this basis.

INTRODUCTION

Simple domain structures, such as stripe domains, CMD, and their combinations, have been under intensive investigation for a long time – about 40 years [1]. First observations of more complex ring-shaped domains were reported by Kandaurova with collaborators in 1992 [2]. They have called this phenomenon "leading center in an anger state" because localized dynamic domain structures of this type arise from chaos under influence of oscillating external magnetic field perpendicular to the film plane. The frequency of oscillations is about 100-1000 Hz what is considerably lower than resonance frequencies of domain walls and their structure elements (horizontal and vertical Bloch lines). According to some experimental works [3,4], related structures having the form of a spiral can remain stable even after the external field is turned off. Analytical description of static stability of the spiral structures can be found in the work [5], but full theoretical explanation for all observed phenomena have not been published yet.

The present work contains an attempt of investigation of ring-domain system's static stability. The first section is devoted to details of the mathematical model. The non-linear integro-differential equation for a magnetization distribution can't be resolved analytically, so a numerical process is developed and realized in a computer program allowing the numerical experiment implementation. The section devoted to results presents the visual method of the system's stability determination and proposes some features related to its dynamical behaviour. The main origin of the system's peculiar properties is highlighted.

THEORY

Mathematical Model

Domains in the thin magnetic film can be represented as regions with the uniform magnetization oriented opposite to the magnetization direction in the rest of the film. Fig.1 shows an example of the simplest structure containing ring-shaped domains.

For the calculation of the domain system's energy local density the well-known expressions for energy components are used [1]. Complete energy includes the anisotropy energy, exchange energy, Zeeman energy of interaction with the external magnetic field, and the energy of

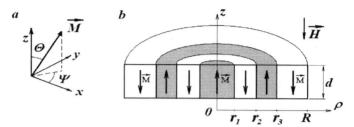

Fig.1 (a) Coordinate systems and the magnetization orientation. The angles Θ
and Ψ determine the magnetization vector $\vec{M}$ direction. (b) A section of the
domain system of CMD-ring type in the cylindrical coordinates (ρ,φ,z). Shaded
regions represent CMD and the ring-shaped domain; d is the thickness of the
magnetic film, r_1 is the CMD radius, r_2 is the internal ring-domain radius, r_3 is the
external radius of the ring, and $\vec{H}$ is the vector of the external magnetic field.

---

magnetostatic dipole-dipole interaction (DDI). The minimization condition for the complete
energy local density functional provides the equation for the function $\Theta(\rho)$ determining the
magnetization radial distribution:

$$\Delta^2 \frac{d^2\Theta}{d\rho^2} + \frac{\Delta^2}{\rho}\frac{d\Theta}{d\rho} - \sin\Theta\cos\Theta - \frac{1}{Q}h_z\sin\Theta + \frac{1}{Q}\sin\Theta\int_0^\infty d\rho'\cdot\rho' K(\rho,\rho')\cos\Theta(\rho') = 0, \quad (1)$$

where $\Delta = \sqrt{A/K_a}$, the domain wall (DW) width parameter, $Q = K_a/2\pi M^2$, the dimensionless
quality factor, K_a, the perpendicular anisotropy constant, and A, the exchange constant, are the
magnetic film parameters; Θ is the angle between the magnetization vector direction and the z-
axis, Ψ, the magnetization azimuth angle, h_z, the z- projection of the normalized external
magnetic field $h = H/4\pi M$, M, the saturation magnetization of the film, and ρ, the polar radius
in the inverted space.

When deriving the equation (1) the system's radial symmetry is implied, so the nucleus of
the magnetostatic dipole-dipole interaction $K(\rho,\rho')$ [5] for the case of $\partial\Theta/\partial\phi = 0$ [6] is:

$$K(\rho,\rho') = \frac{2}{\pi d(\rho+\rho')}\left[K\left(\frac{2\sqrt{\rho\rho'}}{\rho+\rho'}\right) - \frac{1}{\sqrt{1+\frac{d^2}{(\rho+\rho')^2}}} K\left(\frac{2\sqrt{\rho\rho'}}{\sqrt{(\rho+\rho')^2+d^2}}\right) \right] \quad (2)$$

where d is the the magnetic film thickness and $K(k)$ are the complete elliptical integrals of the
first kind.

158

Equation (1) is a non-linear singular integro-differential equation. It's possible to use the quasilinearization method with a start approximation $\Theta^{(0)}(\rho)$ in order to transform it into the linear one. The problem of keeping the system's non-linear features enters into the problem of a start approximation choice. Presumably, a chain of so-called π-kinks representing the well-known solution of the Landau-Lifshitz equation for the isolated DW [7,1] can be used:

$$\Theta^{(0)}(\rho) = 2 \cdot \sum_{i=1}^{N} (-1)^i \, arctg \left\{ exp \left(\frac{\rho - \rho^{(i)}}{2 \cdot \Delta_{st}} \right) \right\}, \tag{3}$$

where $\rho^{(i)}$ determine the kinks locations, N_k is the number of kinks, and Δ_{st} is the DW width parameter of the start approximation. After linearization we obtain:

$$\Delta^2 \frac{d^2\Theta}{d\rho^2} + \frac{\Delta^2}{\rho}\frac{d\Theta}{d\rho} + p(\rho)\Theta - \frac{1}{Q}\sin\Theta^{(0)}\int_0^\infty \rho' K(\rho,\rho')\sin\Theta^{(0)}(\rho')\cdot\Theta(\rho')d\rho' = f(\rho). \tag{4}$$

Exact expressions for functions $p(\rho)$ and $f(\rho)$ are too complicated to write them here in full (see them in [6]). The start approximation (3) curve accompanied with the results of the solution of equation (4) are shown in the fig. 2. The solution procedure is described below.

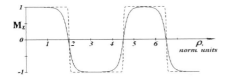

Fig. 2. Example of equation (4) solution results: dotted line – start approximation, solid line – solution result. Domain wall width tunes to the DW width parameter in equation (4) while walls itself are frozen.

Numerical Procedure

It should be noted that all of the variables having the dimension of length could be normalized on film thickness d and the energy on $4\pi^2 M^2 d^3$. All figures below are expressed in these units.

We use a discrete-point representation (DPR) for any function of the polar radius ρ and enumerate points ρ_i of the equidistant n-part partition with index i. The sample's radius R has a large enough value for the magnetization near the outer boundary to be uniform and parallel to the external field direction. We replace the integral in equation (4) with a finite sum using the trapezoidal formula [8] and the derivatives are replaced with the finite difference relations. Equation (4) transforms into the set of linear algebraic equations:

$$\left(\frac{\Delta^2}{s^2} - \frac{\Delta^2}{2s}\frac{1}{\rho_i} \right)\Theta_{i-1} + \left(p_i - \frac{2\Delta^2}{s^2} \right)\Theta_i + \left(\frac{\Delta^2}{s^2} + \frac{\Delta^2}{2s}\frac{1}{\rho_i} \right)\Theta_{i+1} -$$

$$- \frac{s}{Q}\sin\Theta_i^{(0)}\left\{ \frac{1}{2}M_{i,0}\Theta_0 + \sum_{j=1}^{n-1} M_{i,j}\Theta_j + \frac{1}{2}M_{i,n}\Theta_n \right\} = f_i ,$$

$$i = 1...n, \tag{5}$$

where s is the partition step size and the matrix elements $M_{i,j}$ are defined as

$$M_{i,j} = \rho_j K(\rho_i, \rho'_j) \sin \Theta_j^{(0)}.$$

System (5), consisting of the n equations on the $n+2$ unknowns $\{\Theta_{-1}, \Theta_0, ..., \Theta_{n+1}\}$, must be completed with boundary conditions (BCs). We can satisfy to the above condition for the outer boundary R and provide the radial symmetry and smoothness of the magnetization distribution when demanding the derivatives to be zero in the center (left BC) and on the boundary (right BC). In DPR the boundary conditions have the form:

$$\Theta_{-1} = \Theta_1, \quad \Theta_{n+1} = \Theta_{n-1} \tag{6}$$

The complete set of equations (5) (6) are resolved numerically with the use of the Gauss method [8]. Calculations show the relations of the equation matrix elements standing at the main and two neighbour diagonals to the remainder to be as large as *1000*, so the matrix is almost tridiagonal. This occasion provides numerical procedure stability. Really, if there is a small or zero element at the main diagonal, the procedure becomes unstable because of the division by a diagonal element at the every step of the Gauss' algorithm.

RESULTS

Solution of the equation (4) provides minimization of the energy local density only, not minimization of the complete energy (see fig.2). Thus we have to explore the complete energy dependencies on the parameters determining the system's geometry. From this point of view, local minimum of the energy corresponds to the stable state of the system.

The method described above is tested on the problem of the static stability of isolated CMD ($N_k=1$). The results are in a good quantitative agreement [6] with the proven data in works [9,1]. Some issues regarding stability of an isolated ring-domain are also discussed in the paper [6]. In the present work we have focus on the investigation of the simplest LDS consisting of CMD and one ring shaped domain around it ($N_k=3$, see fig. 1).

If the external magnetic field value h is fixed, the energy of CMD-ring system depends on three parameters: CMD radius r_1, internal r_2, and external r_3 radiuses of the ring-domain. So we have to minimize the function of three variables. Moreover, it must be repeated for various values of the field h. For a solution of this problem a non-formal method, allowing to visualise the energy dependencies as surfaces and demonstrate some interesting properties of the system, is chosen.

After the preliminary calculations it has been found that all of the surfaces, received for the fixed value of one of three geometrical parameters, have an obvious area of energy reduction. This area can be called "valley" because its lowest energy value corresponds to the certain fixed width of the ring-domain r_3-r_2. For example, fig.3 shows one of such surfaces. When the value of the external magnetic field increases, the ring-domain width decreases approaching to the zero value asymptotically. Taking into account a general view of the surface, we can deduct that the energy minimum, if it exists, can be located only at the bottom of the valley. Besides that, the existence of the valley allows us to assume that ring-domain width will be conserved even in the case of the system's slow dynamics.

The energy surfaces research highlights another important system peculiarity, namely, the effective interaction of domain walls. This interaction can be deduced from the existence of the energy barriers between internal and external borders of a ring-domain, as well as between CMD wall and the ring wall. Moreover, there is a DW attraction, which can provide system stabilization in the balance with DW repulsion. The presence of such an interaction results in compression of CMD inside the ring and forces it to collapse at a field value smaller than that for an isolated CMD. Next, DW-repulsion hamper the annihilation of the ring-domain walls. Domain of such a form can collapse only after its transformation into CMD [6].

The use of the confinement condition r_3-r_2=const reduces a number of independent parameters down to two if external magnetic field h is fixed. They are r_1 and r_2. Thus all of the energy dependencies can be presented as surfaces (see fig. 4).

A valley's bottom inclination makes the basic distinction between the surfaces constructed for various values of the magnetic field. If the field is great enough (more than 0.25), the valley bottom has the gradient towards to smaller sizes of the ring-domain. Thus there is also the CMD compression and, when r_1 becomes as small as 0.5, the domain collapses. After that, the ring-domain transforms into CMD, which remains stable up to field's value h^k=0.315. Certainly, deviations from the described order of transformations in the system are possible, but their realization is complicated by presence of the potential barriers.

At the weak fields ($h \leq 0.24$) the ring-domain has a tendency for moving away from the center, theoretically at an infinitive distance, while keeping the ring width fixed. Simultaneously, the CMD radius r_1 increases and approaches asymptotically to the value of the isolated CMD radius appropriate to the given value of the external field. It's important to note that the ring-domain can have an influence on the CMD size even when CMD radius r_1 and ring radius r_2 differ more than 10 times in values.

In the field range 0.245-0.247 the valley bottom is near to horizontal, which can correspond to a situation of indifferent stability of the system. This makes it possible for the system to become stable under influence of a rather weak factor not taken into account by the model. For instance, it can be a magnetic film defect of any kind. Sections of energy 3D-data volume of the r_3-r_2=const type, constructed for various values of the ring width, confirm these conclusions, as well as validity of the assumption about fixed value of the ring's width.

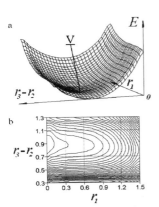

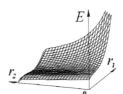

Fig. 4. Energy of a CMD-ring system calculated for r_3-r_2=const. The system shows a tendency to an expansion.

Fig.3. Energy of a CMD-ring system calculated for r_2=const: (a) surface plot, V – "valley" determining the fixed value of the ring-domain width, and (b) contour plot.

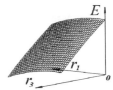

Fig. 5. Energy of a CMD-ring system calculated for r_2=const. Magnetostatic dipole-dipole interaction energy is not included.

Last, but not least, if the energy contribution arising from DDI is eliminated, all of the described above features of the energy surfaces disappear (see fig. 5). Thus all peculiar properties are conditioned by the magnetostatic dipole-dipole interaction.

CONCLUSIONS

The results described above allow us to assume that the magnetostatic stabilization of the CMD-ring system is impossible, but there is a metastable state, which can become a stable one under influence of a rather weak factor not taken into account by the model. Probably, the ring domain can exist as an anger state of the domain system. The energy dependencies on system's geometry parameters have some features that allow us to suppose that in the case of slow dynamics of the system the ring-domain width will be frozen. We conclude that the ring domain couldn't collapse by merging its borders. The process of its appearance-disappearance necessarily goes through the stage of transformation into CMD, and CMD in the center of the system can hamper the ring's disappearance. So CMD can play the role of a "leading center" for formation of the system of several ring-shaped domains if the external field is oscillating, as experimenters have observed it. The magnetostatic dipole-dipole interaction is the necessary condition for existence of all of the system's peculiar properties.

REFERENCES

1. A. P. Malozemoff and J. C. Slonczewski, Magnetic Domain Walls In Bubble Materials, (Academic Press, New York, 1979), pp. 9-159.

2 G. S. Kandaurova and A. E. Sviderskiy, Physica B, **176**, 213 (1992).

3. Yu. L. Gobov and G. A. Shmatov, Fiz. Met. Metalloved, **78** (1), 36 (1994).

4. A. F. Reiderman and Yu. L. Gobov, Defektoskopiya, **3**, 70 (1989).

5. A.B. Borisov and Yu.I. Yalishev, Fiz. Met. Metalloved, **79** (5), 18 (1995).

6. A. F. Gal'tsev and Yu. I. Yalishev, Fiz. Met. Metalloved, **85** (4), 5 (1998).

7. L. D. Landau and E. M. Lifshitz, Phys. Zs. Sovjetunion, **8**, 153 (1935).

8. Any handbook dealing with the numerical methods. We use the next one:
   N. S. Bahvalov, Chislenniye metody, v.**1**, (Nauka, Moscow, 1973) 631p.

9. A. H. Eschenfelder, Magnetic Bubble Technology, (Springer-Verlag, New York, 1981) pp. 35-87.

MATHEMATICAL MODELLING OF ROUGHNESS-TEMPERATURE EVOLUTION

S.NEDELCU, N.MOLDOVAN
Institute of Microtechnology, P.O.Box: 38-160, 72225 Bucharest, Romania, sorinn@imt.ro

ABSTRACT

Orientation -dependent etching of silicon in alkaline solutions is close to get a satisfactory model by considering the random Si-Si bond breaking algorithms (BBA). Here we report our progress regarding the simulation of the atomic-scale roughness of the etched surface, by considering the influence of the processing temperature. We chose the model parameters to fit the experimental data for the etching rate angular dependence for $80^{\circ}C$. We supposed an Arrhenius type dependence on temperature for the basic bond-breaking probabilities and obtained the etching rate and roughness dependence on temperature. The temperature dependence of the etching rates showed the main features of the experimental data: an increase of the etching rate with temperature, a slight change of anisotropy and an increase of the roughness. Further investigations showed we could be close to a roughening transition, which manifests mainly along the directions of the etching rate maxima.

INTRODUCTION

The bond-breaking algorithms for simulating the orientation dependent etching of silicon in alkaline solutions were able to provide angular diagrams close to the experimental ones, both in Monte Carlo [1] and in the master equation formulation [2]. The master equation method showed advantages by operating directly with average values, with the price of complicating the calculations. However, this method permits the simple choise of orientations while preserving the "large area condition", i.e. the fact that the considered etched planes are indefinitely large, while the absence probabilities of bonds laying in the same plane have equal values. That is why we chose this model to continue our studies. Recent developments of the method showed the benefits of solving the discrete master equation by means of considering the fields of continuous extensions of the absence probabilities of bonds (bonds are defined only in certain discrete lattice locations, while the fields are continuous all over the lattice domain). The continuous fields and the large area condition reduce the master equation system to four 1-dimensional equations describing the evolution of four coupled fields [2] $P(z,t)$, $Q(z,t)$, $M(z,t)$, $N(z,t)$, each corresponding to one certain orientation of bonds in the silicon lattice.

THEORY

The parameters entering the master equation are ten basic probabilities P_0, P_1, P_{11}, P_2, P_3, P_{12}, P_{13}, P_{22}, P_5, P_6, of breaking in the unit of time the bonds surrounded by 0 to 6 first order neighbouring Si-Si bonds. After an analysis of the possibility to deduce the values of these probabilities from atomic properties, for assuring the microscopic reversibility condition for the partial reactions, we concluded upon the necessity to introduce the possibility for some bonds to be re-established. Broken bonds can be re-bonded in the time unit with different probabilities, depending also on the state of the surrounding first order neighbours. This introduces ten new "re-bonding" parameters: Q_0, Q_1, Q_{11}, Q_2, Q_3, Q_{12}, Q_{13}, Q_{22}, Q_5, and $Q6$. The completed form of the master equation is given in the appendix.

The P-parameter subspace preserves the significance within the polyhedral structure described in [2], due to the inequations governing the P-parameters. As long as we search for silicon/alkaline etchant type phenomena, we are restricted in a small part of the P-subspace, the vicinity of point B (Fig.1). The re-bonding scenario corresponds to different types of crystal growth. However, as long as we speak about etching, we should be in a certain part of the Q-parameter subspace, defined as $P_{ij} \geq Q_{ij}$. All the Q parameters were described as fractions α_{ij} of the corresponding P parameters ($Q_{ij}=\alpha_{ij}P_{ij}$), where $\alpha_{ij} \leq 1$. If we associate to the Q-subspace a similar structure as for the P parameters, the current point will lie also in the close vicinity of "point B" (Fig.1). The described model is similar to the model for crystal growth described by Gilmer and Bennema [4], being adapted to the problem of etching diamond-like lattice structures. Q_6 – the probability of re-bonding a bond having no first-order neighbouring Si-Si bonds was chosen zero (it is a kind of nucleation of bonds in the middle of the etchant, which is obviously a very unprobable event in praxis). A study of sensitivity of the etching rate diagrams with respect to large changes in the values of the Q-parameters showed that the parameters Q_1, Q_{11}, Q_2, Q_3, Q_{12}, and Q_{13} are unessential (in connection with the high sensitivity with regard to the corresponding P parameters).

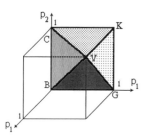

Fig.1. The reduced 3-D parameter space for the usual cases of anisotropy (after [2])

Table I. The reference values of the P and Q parameters at 80°C. $p_{ij}=P_{ij}/P_6$, $\alpha_{ij}=Q_{ij}/P_{ij}$ are dimensionless. The relative activation energies are expressed in eV.

p_0	p_1	p_{11}	p_2	p_3 to p_6
0	2×10^{-6}	2.5×10^{-3}	5×10^{-3}	1

E_{p0}	E_{p1}	E_{p11}	E_{p2}	E_{p3} to E_{p6}
-	0.189	0.183	0.162	0

α_0	α_6	α_5	α_{13}	α_{22}	Other α_{ij}
0	0	1×10^{-2}	0.9	0.5	0.999

E_{q0}, E_{q6}	E_{q5}	E_{q13}	E_{q22}	E_{q3}	E_{q1}, E_{q11}
-	0.14	0.0032	0.021	3.0×10^{-5}	0.4

Since we have in mind to develop a method to extract the values of the P and Q parameters from atomic and molecular properties, the continuation of the simulation experiments were done to verify qualitatively if the predictions of our model so far correspond to practical observations. For this purpose, we chose (by successive trials) the remaining Q parameters (in fact, the α parameters) to make the etching rate diagrams to fit the experimental data obtained on Si/KOH/80°C by Sato et.al. [3]. The values of the P and Q parameters at 80°C (chosen as a reference) are given in Table I.

The basic P and Q parameters are expected to have a temperature dependence of exponential form: $P_{ij}=\nu \exp(-E_{Pij}/kT)$, $Q_{ij}=\nu \exp(E_{Qij}/kT)$, where E_{Pij}, E_{Qij} are activation energies and ν is a frequency-type pre-exponential factor, associated to the material and process [4]. We will not discuss here the degree of correctness of these suppositions, but resume to apply the model of Gilmer and Bennema [4], which provided valuable qualitative results for the crystal growth. Since the frequency ν, like the whole P_6 parameter, can be forced as a common factor in the master equation, it can be made to multiply the time unit, this way obtaining a dimensionless temporal variable. We also resume to describe the activation energies as measured relative to the activation energy of P_6. If we fit the simulated etching rate angular diagram to the experimental one at 80°C, we can get the values of P_{ij} and O_{ij} at that temperature, as well as the relative activation energies E_{Pij}, E_{Qij}. Having these energies, we can calculate the P and Q parameters at any temperature. The relative activation energies calculated this way are also given in Table I.

For verifying the temperature dependencies, we chose two significant outputs of the simulation program: the etching rates and the roughness. The etching rate v is obtained by deriving the average depth of etching h_{ma} with respect to time:

$$h_{ma} = \int_0^\infty z \frac{dP_a}{dz} dz, \qquad v = \frac{dh_{ma}}{dt} \qquad (1)$$

where $P_a(z,t)$ is the absence probability of an atom located at the depth z, at the moment t, obtained as the product of the absence probabilities of the surrounding bonds.

The roughness of the etched surface was calculated as the standard deviation of the etching depth:

$$Rf = \sqrt{\int_0^\infty (z^2 - h_{ma}^2) \frac{dP_a}{dz} dz} \qquad (2)$$

Both the etching rate and the roughness can be represented as functions of the etching time or as functions of the etching depth (via the etching depth – time dependence).

SIMULATION RESULTS

In all the calculation results, the etching rates are expressed in units equal to a x P_6, where a is the lattice constant (a=5.43Å for Si) and P_6 contains the time unit. Since we are interested not explicitly in the absolute value of one etching rate, but in the shape of the angular diagram, we can keep P_6 undefined (arbitrary units). The roughness is expressed in units equal to a.

We performed simulations for temperatures between 58°C and 95°C. The changes in the etching rate diagrams for that temperature range, are depicted in Fig. 2 for the (hk0) and (hkk) families of planes. Fig.3 presents the Arrhenius plots of the etching rates for the (110), (100), (211) and (111) orientations, calculated for 1°C step-width. The roughness diagrams (in Cartesian coordinates) for the same temperatures are presented in Fig.4. Fig.5 shows the temperature dependence of the roughness for the same orientations.

DISCUSSIONS

The etching rates generally increase with temperature. However, for the maxims of the etching rates around the (250) direction (70°), the etching rate for 90°C is higher than for 95°C. The increase in temperature seems to reduce the diagram's protrusion around the <100> family of directions (Fig.2, left). The changes are much smaller in the (hkk) diagram (Fig.2, right).

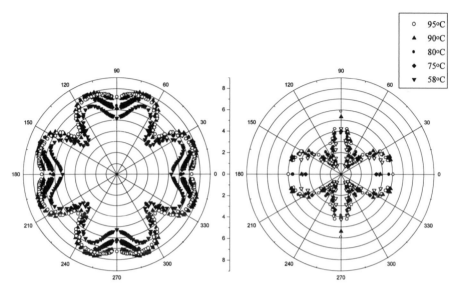

Fig.2. Simulated etching rate angular diagrams for the <hk0> (left) and for the <hkk> (right) family of directions, for several etching temperatures between 58°C and 95°C. The angular origin (0°) of the diagrams correspond to the <100> directions.

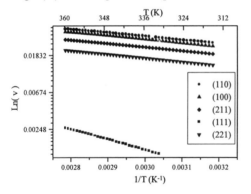

Fig.3. Arrhenius plots of the etching rates for several directions, between 40°C and 85°C. The lines corresponding to the directions <111> and <110> cross around 464°C.

●	(110)
▲	(100)
◆	(211)
■	(111)
▼	(221)

Some characteristic features of the temperature dependence of the etching rates for several directions can be observed in the Arrhenius plots (Fig.3). The general tendency is to reduce anisotropy with increasing temperature, in agreement with the experimental data of Seidel et. al. [5]. However, not all the lines cross at the same temperature, some of them seeming rather parallel than crossing each other.

One characteristic feature observed in the temporal evolution of the low-index planes is a periodic change of the etching rate, in correlation with the succession of atomic planes. The etching rates figured in the plots of Fig. 2 and 3 correspond to the maximal encountered values of those etching rates. These values provide smooth continuations of the rest of the diagrams.

A special remark has to be done regarding the calculated roughness values. For low temperatures (under 85°C), the values of the roughness for all directions show a saturation with time or etching depth. At 90°C, the fast etching directions show a continuous and constant growth of the roughness. This corresponds to a roughening transition for those directions, as described by Gilmer and Bennema [4]. The roughness values figured for 90°C and 95°C in Fig.4 correspond to the maximal values of roughness reached at the end of the computation cycle and are not saturation values. However, the roughening transition occurs not for all directions at the same temperature, which keeps part of the points spread along smooth curves. The roughening transition is a consequence of the growth part of the master equation (the Q-parameters). A practical verification of this tendency will probably be hard to fulfil, since the transition temperature is suggested to lie close to or above the boiling point of the etchant.

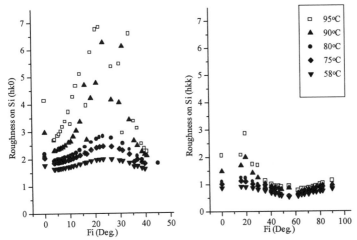

Fig.4. Cartesian plots of the angular dependence of roughness for the <hk0> (left) and <hkk> (right) directions of etching, for several temperatures. The roughness values are expressed in lattice constants (a-units).

The temperature dependence of roughness seems to be quite irregular (Fig.5). The general tendency is to grow with temperature, however, in different ways. The (111) planes of etching encounter the most stable roughness, which are almost independent on temperature.

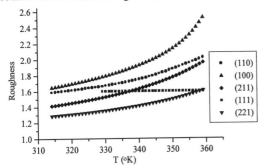

Fig.5. The roughness dependency on temperature for some low-index etching planes, between 40°C and 85°C, bellow the roughening transition temperature. The roughness values are given in lattice constants (a).

CONCLUSIONS

We extended the bond breaking scheme based on the master equation algorithm to include the local re-bonding during wet anisotropic etching of silicon, in order to assure the microscopic reversibility of the considered reactions. A proper choice of the model parameters to fit the experimental results at 80°C provided the possibility to simulate the etching rate diagrams and the roughness behaviour with temperature. Our model reproduces the main features observed in practice –the decrease of anisotropy with temperature, the activation of the etching rates and the increase of roughness with temperature. The etching process seems to reach a roughening transition at temperatures close to the boiling point of the etchant. Above this temperature, part of the etching directions show a constant growth of the roughness in time. However, some of the etching rates reach saturation values even above this temperature, without showing any discontinuity around the transition temperature.

ANNEX

The master equation system for the completed etching - rebinding scheme can be written using the paranthesis-opperator $\{,,,,,\}$ defined in [2] for the P-parameters. A similar operator, designated as $\{,,,,,\}_Q$ can be defined simply by using instead the P parameters, the Q parameters. For regularity reasons, the parenthesis-operator using the P-parameters will be designated as $\{,,,,,\}_P$. With these notations, the completed master equation system, transformed to 1- dimension (z) on basis of the large area condition, has the form:

$$\frac{dP}{dt} = P\{Q(z+R/2,t),M(z+S/2,t),N(z+T/2,t),Q(z-R/2,t),M(z-S/2,t),N(z-T/2,t)\}_Q -$$

$$(1-P)\{Q(z+R/2,t),M(z+S/2,t),N(z+T/2,t),Q(z-R/2,t),M(z-S/2,t),N(z-T/2,t)\}_P$$

$$\frac{dQ}{dt} = Q\{P(z-R/2,t),M(z-(R-S)/2),N(z-(R-T)/2),P(z+R/2),M(z+(R-S)/2),N(z-R-T)/2)\}_Q -$$

$$(1-Q)\{P(z-R/2,t),M(z-(R-S)/2),N(z-(R-T)/2),P(z+R/2),M(z+(R-S)/2),N(z-R-T)/2)\}_P$$

$$\frac{dM}{dt} = M\{P(z-S/2), Q(z-(S-R)/2), N(z-(S-T)/2), P(z+S/2), Q(z+(S-R)/2, N(z+(S-T)/2)\}_Q -$$

$$(1-M)\{P(z-S/2), Q(z-(S-R)/2), N(z-(S-T)/2), P(z+S/2), Q(z+(S-R)/2, N(z+(S-T)/2)\}_P$$

$$\frac{dN}{dt} = N\{P(z-T/2), Q(z-(T-R)/2), M(z-(T-S)/2, P(z+T/2), Q(z+(T-R)/2), M(z+(T-S)/2)\}_Q -$$

$$(1-N)\{P(z-T/2), Q(z-(T-R)/2), M(z-(T-S)/2, P(z+T/2), Q(z+(T-R)/2), M(z+(T-S)/2)\}_P$$

where $P(z.t)$, $Q(z.t)$, $M(z.t)$, $N(z.t)$ are the absence probabilities of the four type of bonds (differing by orientation), $R=l+m$, $S=m+p$, and $T=l+p$ and l,m,p are the Miller indexes of the etched plane. The Q-parameters and the absence probability $Q(z,t)$ should not be confused.

REFERENCES

[1]-H.Camon, N.Moldovan, Z.Moktadir, M.Djafari-Rouhani, M.Ilie, S.Nedelcu, ECOSS 17, Genoa, Italy, Sept.1996
[2]-N.Moldovan, S.Nedelcu, H.Camon, MRS Fall Meeting, Boston, 1997, paper Q5.2
[3]-K.Sato, A.Koide, S.Tanaka, Proc. of J.F.F.F. Symp., Japan, p.19-27, 1989
[4]-G.H.Gilmer, P.Bennema, J.Appl. Phys., Vol. 43, No.4, April 1972
[5]-H.Seidel, L.Csepregi, A.Heuberger, H.Baumgärtel, Journal of Electrochem.Soc.Vol.137, Nr.11, Nov.1990, p.3612

SELF-ORGANIZATION OF FULLERENE CLUSTERS

SLAVA V. ROTKIN
Ioffe Physico-technical Institute, 26, Politekhnicheskaya st., 194021, St. Petersburg, Russia
E-mail: rotkin@theory.ioffe.rssi.ru

ABSTRACT

The paper introduces the plasmon-Frenkel-exciton model [1] for the fullerene solids and explains a possible reason for the formation of the van-der-Waals [2] C_{60} cluster complexes. The interaction of the cluster and the organic (non-polar) liquid solvent will be considered. The motivation of this interest is related also to the experiments on the photoclusterization in the water solution [4] and cluster formation in beams [5].

ENERGY OF VAN-DER-WAALS INTERACTION

Typically van-der-Waals energy is given by shift of plasmon zero-oscillation energy in a solid comparing with a single cluster. This shift is due to Coulomb interaction between clusters. The origin of the van-der-Waals forces is the same as the origin of depolarization shift of dipole frequency in a dielectric medium. One can easily calculate this depolarization shift following the method of the mean field for the case of cubic crystal and for rotationally invariant system (liquid, for example). That is valid for fullerene condensed matter, which forms fortunately face-centered cubic lattice with four clusters in a cell.

Let us calculate the mode frequencies taking into account dipole-dipole interaction. It was shown that higher multipole interaction terms can be neglected [6].

The dipole excitation of a single cluster is a basic unit in our consideration. An electron density of C_{60} is known to possess a collective mode with the frequency about 25 eV. This is the surface plasmon on the fullerene sphere. It was obtained within phenomenological model [6a] as well as within more sophisticated approach (see [7] and references [10-14] in Ref.[7]). We will use a dipole plasmon as an elementary excitation of the fullerene "super-atom" unit.

The van-der-Waals interaction can be written using the fluctuation-dissipation theorem as an integral over the frequency of combination of dynamic polarizability of the C_{60} cluster and dielectric function of the medium. Then this integral can be evaluated in the complex plane of the frequency. The collective plasma mode of the cluster, having the maximal frequency between dipole excitations, makes the main contribution as an excitation having the maximal oscillator strength.

Let us remind that the frequency of the multipole (in particular, dipole) collective mode of C_{60} coincides with plasmon frequency of a hollow metal sphere [11]:

$$\omega_1 = \omega_p \sqrt{\frac{2}{3}} \sim 22 \text{ eV} \qquad \omega_p = \sqrt{\frac{240 \, e^2}{mR^3}} \sim 26.9 \text{ eV}, \tag{1}$$

here ω_1 is the dipole mode frequency, ω_p is the plasma frequency, 240 is the number of valence electrons of the cluster, m an e are electron mass and charge. The sphere radius R is taken ~ 3.3 Å to describe the fullerene plasmon properly. Considering the fullerene solid we use the Lorentz-Lorenz approximation basing on a high polarizability of the single cluster, α, which reads as:

$$\alpha(\omega) = R^3 \frac{1}{1 - \omega^2/\omega_1^2}. \tag{2}$$

We took $\alpha(0) \simeq R^3$ which is valid with a high accuracy [8]. A packing factor, coming into the dielectric function along with the dynamic polarizability, is as follows:

$$\eta = 4\pi\nu\alpha(0) = 4\pi\frac{4}{d^3}R^3, \tag{3}$$

where $\nu = 4/d^3$ is the density of the clusters, $d \sim 14.2$ Å is the lattice constant. This is very common parameter in such a calculation, we take it as 0.79. It will be very convenient to use as a dynamic variable the square of dimensionless plasma frequency:

$$x = \omega^2/\omega_1^2.$$

Then the Lorentz-Lorenz formula gives the high frequency limit of dielectric function:

$$\epsilon(\omega) = \frac{1 + 2/3\eta/(1-x)}{1 - 1/3\eta/(1-x)} = \frac{1 + 2/3\eta - x}{1 - 1/3\eta - x} = \frac{x_L - x}{x_T - x} \tag{4}$$

which in turn gives us plasma longitudinal frequency $\omega = \omega_1\sqrt{x_L} \sim 26.2$ eV along with transverse excitation frequency $\omega = \omega_1\sqrt{x_T} \sim 19.6$ eV as zero and pole of the ϵ, the dielectric function. Then the van-der-Waals energy is simple difference between bare plasma frequency of the cluster and the modes in the solid. It reads as:

$$W = \hbar\omega_1\left(\sqrt{x_L} + 2\sqrt{x_T} - 3\right) \sim -\hbar\omega_1\frac{\eta^2}{24} \tag{5}$$

The only two parameters determining the van-der-Waals energy are the plasmon energy and the packing factor. In the expression above we used the expansion on the small η. The interplay between these parameters gives us, for example, the plasma frequency on the boundary between some mediums, in the liquid, in the medium with polarizable dopant and so on. We will discuss it at length elsewhere. Substituting the numbers into Eq.(5) one gets the van-der-Waals energy about –1.1 eV per cluster in the solid. The one of the first papers containing similar consideration to be mentioned is [8]. In the next section we will compare the result with the energy in solution.

FULLERENE IN SOLUTION

The plasma frequency in a solution is lower than in a solid phase owing to the depolarization shift is much weaker in any typical organic solvent than in a fullerene solid. The reason is that the fullerene cluster has the very high frequency of this bare plasmon due to large number of highly polarizable electrons. A standard medium is nearly transparent at this frequency. More precisely the dielectric function of the medium is slightly less than the unity at the frequency of C_{60} plasmon.

Below we present a correct method of calculation of the frequency of plasma mode of C_{60} in a liquid insulator. According [7], the surface plasmon in C_{60} is a spherical oscillation of electron density σ_{LM}. For central symmetry of the cluster we use expansion of all quantities in complete spherical harmonics $P_L(r)Y_{L,M}(\Omega)$ those form a complete set on a sphere. In spherical geometry a radial jump in electric field is given by:

$$\frac{2L + 1}{R}\varphi_{LM}^{\text{ind}} = 4\pi\sigma_{LM} \tag{6}$$

where φ^{ind} is an induced part of potential, L, M are multipole power indexes. We close the equation system by writing the response function for fullerene cluster as:

$$\frac{4\pi R}{2L + 1}\sigma_{LM} = -\frac{4\pi R}{2L + 1}\chi_L \varphi_{LM}^{act} = \simeq \frac{\omega_L^2}{\omega^2}\varphi_{LM}^{act}, \tag{7}$$

where $\varphi^{act} = \varphi^{ext} + \varphi^{ind} = \varphi^{ext}_{LM} + 4\pi R\sigma_{LM}/(2L+1) + \varphi^{sol}_{LM}$ is an acting potential, L, M are multipole power indexes. The selfconsistency of the calculation is proved by using of this acting potential, including an induced potential of C_{60} plasmon as well as the potential occurring owing to charge density induced in the solvent. Here χ is a response function of single sphere. This expression is easily obtained from classic charge liquid equations [6a,7]. This consideration is more general than our $L = 1$ case of the dipole plasmon mode. However for the spherical symmetry the expression Eq.(7) holds for any multipole.

When a potential induced in the solvent is absent, $\varphi^{sol}_{LM} = 0$, we return to bare plasmon frequency. The corresponding bare dispersion equation reads as:

$$-\frac{4\pi R}{3}\chi_1 = \frac{\omega_1^2}{\omega^2} = 1 \quad \text{or} \quad \frac{1}{x} = 1. \tag{8}$$

We will simply change the unity to the dielectric function to take into account the solvent potential. It is easily seen substituting Eq.(7) into Eq.(6) and taking the standard RPA sum. As a result the plasma frequency in solution is:

$$\Omega(\varepsilon) = \omega_1\sqrt{\varepsilon(\Omega)} \tag{9}$$

The frequency is smaller than bare C_{60} frequency. So far we obtain the plasmon in the fullerene solid and in the solution. We will use for the dielectric function of the liquid solvent the common formula:

$$\epsilon(\Omega) = \frac{\Omega^2 - \omega_L^2}{\Omega^2 - \omega_T^2} \tag{10}$$

where ω_L is a typical longitudinal frequency of the $\epsilon(\Omega)$, and ω_T is a transverse frequency. With this definition the van-der-Waals energy can be written as:

$$W \simeq -3/2\hbar\omega_1 \frac{\omega_L^2 - \omega_T^2}{\omega_T^2} \frac{\omega_T^2}{\omega_1^2} = -3/2\hbar\omega_1 (\epsilon_0 - 1) \left(\frac{\omega_T}{\omega_1}\right)^2 \tag{11}$$

here ϵ_0 is the static permittivity of the solvent, which is related to the transverse and longitudinal frequencies. The typical values for ϵ_0 is 2.3 for the benzene, 2.4 for the toluene. The van-der-Waals energy is about -0.2 eV for these solvents. So we have to conclude that by this energetical reason the solid fullerene should be more stable.

Let us consider fullerene dimer, the similar problem was done in Ref.[6] for C_{119} molecule. The plasmon frequency is split in axial field. Therefore new modes bring the energy of interaction between clusters in the dimer. Then the van-der-Waals energy of the dimer coupling reads as:

$$W \simeq -3/2\hbar\omega_1 \left(\frac{R}{H}\right)^3 \left(1 + \frac{3}{2}\frac{\omega_L^2 - \omega_T^2}{\omega_T^2}\frac{\omega_T^2}{\omega_1^2}\right) \tag{12}$$

where $H \simeq 8$ Å is the inter-cluster distance; here the last term comes from the solvent depolarization, it is a small correction (about a percent) which will be neglected. The typical value of the van-der-Waals energy of such a dimer is about -0.7 eV.

PLASMON MODES IN ICOSAHEDRAL COMPLEX

Recent experiment shows that in fullerene beams the supercluster is forming [5]. The mass-spectra analysis signs that the cluster of 13 C_{60} is especially stable. The most evident structure of such super-cluster is an icosahedra made by central fullerene and closed packed all other 12 clusters. Our aim is to find a solution for self-consistent high-frequency polarizability of the supercluster and calculate the van-der-Waals forces. The poles of polarizability as before are given by the frequencies of plasmon modes in $(C_{60})_{13}$. We will briefly discuss the method of calculation of these frequencies for icosahedral system.

First of all, one can account a fullerene-fullerene interaction to evaluate the shift of plasmon frequency in a supercluster. It was shown [6] that for fullerene dimer the dipole-dipole interaction is enough to be taken into account (all higher multipole corrections are negligible). The dipole approximation will be used below. The first shift of dipole plasmon frequency of the central C_{60} is given as a sum of fields from all other external fullerenes. The same time the field of the central cluster influences on the external ones. Besides that each external C_{60} has five more nearest neighbors acting on it. To account the mentioned above terms in average one has to solve the dynamical matrix of dimension 39×39. Each dipole plasmon has 3 components (which are in general not only shifted but split).

The full solution of the problem will be given elsewhere. Let us show here the mean-dipole-moment approximation of the problem which allows the analytical solution and allows to evaluate the amount of the van-der-Waals energy. The exact way to do this approximation comes from the group-theoretical consideration.

The 13 fullerenes of the supercluster form Γ_0, the reducible representation (RR) of the icosahedral group Y. The dipole problem stated above has the Hamiltonian (dynamical) matrix forming another RR of Y group which is the direct product of dipole representation and Γ_0. It can be expanded into the direct sum of irreducible representation (IR) of Y:

$$T_1 \times \Gamma_0 = A + 4T_1 + T_2 + 2G + 3H \qquad (13)$$

where we use the standard notation A, T, G, H for IRs of Y with the dimensionality $1, 3, 4, 5$. The T IR has an additional label because of there are two distinct IRs of such type in Y. Simple check of the overall dimensionality judges that the expansion is not false: $1 + 12 + 3 + 8 + 15 = 39$. Note, that this RR contains only 3 dipole active modes. What are them? If one considers the RR of single central cluster, one finds only one icosahedral IR, and it is exactly of such a type. The second IR comes from the spherically averaged 12-cluster shell of external fullerenes. The last harmonic, having full icosahedral symmetry has more complicated symmetry and will be discussed elsewhere. Let us write the dynamical matrix in the following form:

$$\begin{vmatrix} x - 1 & \mathcal{T}_C \\ \mathcal{T}_C & x - 1 + \mathcal{T}_B \end{vmatrix} \qquad (14)$$

where the submatrices $\mathcal{T}_C$ and $\mathcal{T}_B$ depict the dipole-dipole interaction of central fullerene with the shell and the interaction within the shell correspondingly. Each C_{60} interacts with five closest clusters from the shell and the central one. The self-energy term $x - 1 \equiv \omega^2/\omega_1^2 - 1$ is as usual given in units of the plasmon energy of the single fullerene ω_1^2. Without interaction this term gives the bare plasma frequency for each dipole plasmon mode, which will be 6-ly degenerate in this case. Let describe more precisely the submatrices of the interaction. It is connected with the dipole-dipole interaction tensor: $\tau_{ij} = \frac{1}{R^3}(\delta_{ij} - 3\, e_i\, e_j)$, where R is the distance between dipoles and $\mathbf{e} = \frac{\mathbf{R}}{|R|}$. Then the interaction terms are given by: $\mathcal{T}_C = \sum_{g \in Y} \tau(\mathbf{g})$ and $\mathcal{T}_B = \sum_{f \in NNA} \tau(\mathbf{f})$. The resulting matrix gives us all plasma mode frequency shifts.

The result is:

$$W \simeq -1/245 \frac{e^2}{R}, \left(\frac{R}{L}\right)^3 \qquad (15)$$

here we evaluate the plasmon dipole momentum as eR. L is the distance from the center of supercluster to the center of C_{60} in the shell, it could be about 10Å. Van-der-Waals energy of such cluster is about 3.1 eV.

172

In summary, we used a plasmon mode approximation for calculation of van-der-Waals energies of different cluster systems. The approach is shown to be consistent for cluster–medium and cluster–cluster interaction. We applied the method and found van-der-Waals energy of formation of $C_{60 \times 13}$ supercluster.

ACKNOWLEDGMENTS

This work was partially supported by INTAS Grant 94-1172, project no. 98062 of Russian program 'Fullerenes and Atomic Clusters' and RFBR Grant 96-02-17926.

REFERENCES

1. V.V.Rotkin, R.A.Suris, Recent Advances in Chemistry and Physics of Fullerenes and Related Materials, Eds. K.M.Kadish, R.S.Ruoff. Pennington, 1997, v.V, pp.147-154.
2. V.V.Rotkin, MRS-93-Fall Meeting, Boston, USA, 1993. 3. V.V.Rotkin, Nanostructures: Physics and Technology'97, St.Petersburg, Russia, pp.335-338, 1997. 4. E.V. Skokan, private communication; Moscow State University, Moscow, Russia.
5. T.P.Martin et al, Phys.Rev.Lett. 70, 20, 3079-3082,1993. 6. V.V.Rotkin, R.A.Suris, Sol.State Comm., v. 97, N 3, 183-186, 1995; 6a. V.V.Rotkin, R.A.Suris, Proc.of International Symposium "Nanostructures: Physics and Technology-95", pp.210-213, St.Petersburg, Russia, 26-30 June 1995.
7. V.V.Rotkin, R.A.Suris, Proc. Symposium on Recent Advances in Chemistry and Physics of Fullerenes and Related Materials, Eds. K.M.Kadish, R.S.Ruoff. Pennington, 1996, p.940-959.
8. V.V.Rotkin, R.A.Suris, Sov.- Solid State Physics, **36**, 12, 1899-1905, 1994.
9. A.A.Lucas, G.Gensterblum, J.J.Pireaux, P.A.Thiry, R.Caudano, J.-P.Vigneron, Ph.Lambin, W.Kratschmer, Phys.Rev.**B 45**, 13694 (92).
10. G. Barton, C. Eberlein, J. Chem.Phys. **95**, N 3, 1512-1517 (1991).
11. V.V.Rotkin, R.A.Suris, Sov.- Solid State Physics, N 5, (in press) 1998.

ENERGETICS OF FULLERENE CLUSTERS

SLAVA V. ROTKIN, ROBERT A. SURIS
Ioffe Physico-Technical Institute, Politehnicheskaya 26, 194021 St.Petersburg, Russia
E-mail: rotkin@theory.ioffe.rssi.ru

ABSTRACT

A new phenomenological model for calculation of formation energy of carbon nano-clusters of definite shape is proposed. The model uses only three energetic parameters: two first, E_c and $\mathcal{E}_5$, being determined from comparison with experimental data, results of computer simulation for various carbon nano clusters, and the last one is dangling bond energy, E_b. Energies of formation of carbon clusters shaped as cylinder, sphere, icosahedral polyhedra, capsule, were calculated in frame of unified phenomenological approach, which allows to judge relative energetical stability of these clusters.

INTRODUCTION

Last years a number of groups obtained various results on the energetics of carbon nano clusters[1-8]. We showed that instead of quantum-chemical computation our 3-parameter model allows to evaluate formation energies of carbon clusters with curved surface. Taken once from experiment or independent calculation, the 3 parameters let us write the formula for an energy of sp^2 carbon atom on a curved surface of very general shape. The formula is applied to infinite and finite nanotube, to spheroidal cluster and to capsule.

The model predicts relative stability of cluster of definite form in respect with others considered. It allows to find the optimal shape of the cluster, which minimizes the formation energy, and to find most stable cluster with any fixed number of carbon atoms.

We compared our calculation with more precise and also time-consuming data. Despite of different computation approaches and certain difference in calculated quantities the computer modeling data corroborates that some generalization of cluster formation theory is possible. This paper is devoted to give first and simplest example of such a phenomenological approach.

ENERGY OF CURVED GRAPHITE-LIKE SURFACE

Carbon clusters with graphite-like lattice will be considered below. Each carbon atom of such cluster has 3 chemical bonds like a carbon atom in graphite. We believe that energy of any carbon nanocluster can be empirically calculated since it is depending on a few geometrical shape parameters.

We will find energy depending on radius of infinite tube first (as a simplest example of graphite-like carbon nanocluster). We suppose that all carbons of infinite tube are placed on a surface of a round-based cylinder and arranged in regular hexagons. That means that we neglect possible relaxation of bonds when an atom displaced inside or outside of regular surface and for simplicity we not include change in bond length too. We will discuss these assumptions elsewhere.

To calculate an additional energy of infinite tube comparing with the planar graphite sheet of the same number of atoms we suppose that an energy of curved surface depends on a squared curvature. We refer this additional curvature energy to each bond: $E_{bond}^{curv} = E_c\,\theta^2$, where E_c is the first phenomenological parameter of the model; in a stick-and-ball picture θ

is an angle between bond direction and its (not scrolled) planar original position (see Figure 1). θ is equal zero in conventional planar graphite. In a limit of small angle $\theta \simeq b/R$, derived from bond length, b, and curvature radius, R. Throughout the paper we will use dimensionless length in units of bond length, $b \simeq 1.4$ Å taken roughly about the value in graphite.

Fig.1. In a stick-and-ball picture a graphite atom lies on the geometrical surface of cylinder. Then the carbon-carbon bonds are declined from original planar positions.

Note that dimensionless curvature of body, k, simply coincides with θ determined above. Next point is that the curvature along the cylinder axis is zero and it is $1/R$ along the perimeter. So one substitutes in equation for curvature energy the effective curvature k_σ along the σ-bond direction. The bond makes angle φ with cylinder guide. It seems us very natural to suppose that angles between 3 σ-bonds are equal to $\frac{2\pi}{3}$ like for graphite. In that assumption the curvature energy per atom is given by simple summation over σ-bond directions. The result is some kind of an invariant in the case of C_3 atomic bond symmetry:

$$\frac{E^{\text{atom}}}{E_c} = \sum_{\sigma_{1,2,3}} k_\sigma^2 = \frac{9K^2}{2} - \frac{H}{2} = \frac{3(k_1 + k_2)^2}{4} + \frac{3(k_1 - k_2)^2}{8} \qquad (1)$$

here we write it in terms of usual principal curvatures of the surface k_1, k_2 and its combinations: Gauss curvature $H = k_1 k_2$ and average curvature $K = (k_1 + k_2)/2$. All curvatures, in general, are functions of a point. We will consider, for the sake of clarity, only surfaces with constant curvature like a cylinder, sphere, plane or some connected parts of these bodies. We make use of continuous approximation of slightly curved graphite-like surface to change the summation over all atoms with the integration. In this case the integration over the cluster surface of the curvature energy per atom gives us the cluster specific area multiplied by E^{atom}. Note that graphite unit cell has an area $\frac{3\sqrt{3}}{2}$ in units of b^2 and possesses two carbon atoms. For a cylinder of radius R it gives:

$$E_{\text{tube}} = \frac{2S}{\frac{3\sqrt{3}}{2}} E_c \frac{9}{8R^2} = \pi\sqrt{3}\, E_c \frac{L}{R} \qquad (2)$$

where the area is $S = 2\pi LR$, L is the tube length. Actually, a direct computation of E_c is much beyond this consideration. We found E_c parameter to be about 0.9 eV fitting our model to specific energy per unit length of an infinite tube taken from the computer simulation[4]. Within our model we get that any local orientation of σ-bond relatively to guides (or axes) of the tube is energetically equivalent, that depends only on assumption of C_3 symmetry of each carbon atom σ-orbitals with respect to its neighborhood. We conclude that this prediction fits to data from[4] quite well (see Ref.[9,10]).

Let us consider an energy related to dangling bonds which are on the open perimeter of carbon nanocluster. Energy of dangling bond for graphite is well known[4,6] $E_b \simeq 2.36$ eV. We argue that, in average over nanocluster, it is close to the value in planar graphite. The total dangling bond energy per cluster is proportional to number of atoms on the perimeter $\mathcal{P}$: $2\frac{\mathcal{P}}{\sqrt{3}}\zeta E_b$, where one has to include a geometrical multiplier ζ. It reflects that the concrete cluster has, in principle, different number of dangling bonds along the geometrical curve defined as perimeter. For example, let consider different types of tubes: "zigzag" tube (cf.[8]) has ζ equal 1, and ζ is equal $2/\sqrt{3}$ for "armchair" type.

At larger R, typically this dangling bond energy grows compensating the decrease of curvature energy. So it could be a minimum of energy for cluster of fixed number of atoms. We discuss it at length in[11].

SPHERE AND POLYHEDRA SHAPE COMPARISON

We are using in our model only 3 phenomenological parameters. We need above defined two for description of cylindrical tube. In this section we will consider spherical fullerenes. A new term in formation energy will appear. It relates to pentagon cells on a surface of spheroidal carbon nanocluster. According to Gauss-Bonnet theorem any closed spheroid does have 12 pentagons, excepting arbitrary number of hexagons. One expects an energy of bond, belonging to pentagon, to differ (likely, to be larger) than for hexagon. Instead of calculating it exactly, we treat it as phenomenological parameter.

One has no direct proof, but as a rule pentagons try to lie as far each from other as possible. It results from school geometry that 12 centers placed at maximal distance on sphere are located in vertexes of regular icosahedron. When a real cluster has Y_h symmetry (truncated icosahedron), like a C_{60} fullerene, it is a semiregular deltahedron. We will also present formation energy for such type of ball-like fullerene in the end of this section.

Now we are going to calculate the energy of a perfect sphere, having the carbon-like lattice on the surface. The curvature of this body is constant and equal to $3/R^2$ per atom with 3 equal bonds. Because of the curvature energy is proportional to the curved surface which, in turn, is proportional to R^2, finally, this term is independent on the sphere radius R. However, this calculation overcount the number of bonds belonging to hexagons, so that 120 surplus bonds should be excluded. As a result the sphere energy reads as:

$$E_{\text{sph}} = \left(\mathcal{E}_5 + \frac{16\pi E_c}{3}\right) - \frac{N_s}{N}E_c \tag{3}$$

where $\mathcal{E}_5$ is the pentagonal defect energy and we introduce a parameter $N_s = 2 \times 60 \times 16\pi/3\sqrt{3}$. It arises when one considers how much it costs to scroll the graphite plane into spheroidal cluster.

Let consider now a spheroidal cluster with non-uniform curvature. Namely, we will consider cluster constructed from planar truncated triangles (deltagons). Following to Ref.[6,7], we make so-called Holdberg polyhedron which consists of 20 deltagonal faces as a simple icosahedron but each face is combined from hexagons. In a vertex a pentagon is placed. The structure can be arranged in two ways. Depending on that, a number of atoms is given by $20 \times n^2$ or $60 \times n^2$ formula, where n is natural. We will discuss here only the second type owing to similarity in calculation. A degree n determines number of hexagon layers between two pentagons, from other point of view it corresponds to an average radius of cluster. All surface curvature is concentrated in edges, excepting a global topological curvature included in 12 pentagonal vertexes according to Gauss-Bonnet theorem. Note that the polyhedron geometry fixes the curvature degree both at vertexes and on edges. These quantities are expressed through "golden section" $\tau = (1 + \sqrt{5})/2$ and amount: $\xi_{vert}^2 = \frac{9}{16}(2\tau)^4\left(\frac{3}{4} - \frac{1}{4\tau^2}\right)\left(\frac{1}{3\tau^2} - \frac{1}{4}\right)$

177

for vertex and $\xi^2_{edge} = 1 - \frac{1}{4\left(1-\frac{1}{4\tau^2}\right)}$ for edge. The corresponding curvatures are about 5 and 3 Å.

The spheroid has the additional term $\mathcal{E}_5$ which comes from pentagon formation as before. Owing to the number of "curved" bonds on edges scales as a length and other terms in polyhedron energy have no any dependence, total energy is linear on n (or equivalently on R, average radius of polyhedron):

$$E_{CNC} = \left(\mathcal{E}_5 + 60\, E_c \xi^2_{vert} + 60\,(n-1)\, E_c \xi^2_{edge}\right) \tag{4}$$

Depending on polyhedron degree n, the specific energy depends on R, average radius of spheroid, determined as $N = 60 \times n^2 = \frac{16\pi R^2}{3\sqrt{3}}$.

It seems to be clear that cluster of infinitely large radius prefers to be shaped uniformly as a sphere rather than to have sharp edge. We replot data from Ref.[4,7] on inverse number of atoms in order to compare model calculation for spherical and polyhedral shape clusters (Fig. 2). For all data we find our results fit quite well except energy of very high mass cluster $N = 960$ that can be beyond the calculation accuracy.

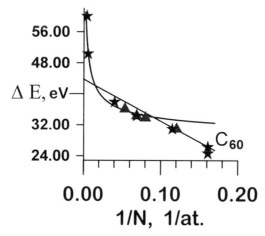

Fig.2. Data from[4,7] agrees with the model calculation showing the relative stability of faceted polyhedron in respect with the spherical cluster in the central region.

We made a comparison of spherical and polyhedral shape cluster energy under equal number of atoms. Surprisingly we found that energy difference between sphere and polyhedron changes sign. It means that there is a cluster mass region, such as the formation of Holdberg polyhedron of order n is energetically favorable: $1 < n < 4$.

OPTIMAL CAPSULE

Here we will consider the cluster which is cylindrical part and two semispheres on the ends of cylinder, so called "carbon capsule." The energy of such cluster reads as:

$$\begin{aligned} E &= E_5 + E_c \left(\frac{16\pi}{3} - \frac{120}{R^2} + \frac{\pi\sqrt{3}}{R}H - \frac{\eta}{R}\right) = \\ &= E_5 + E_c \left(\frac{16\pi}{3} - \frac{N_s}{N_c} - \frac{\eta}{R} + \frac{9}{8R^2}(N - N_c)\right) \end{aligned} \tag{5}$$

where one easily recognizes terms coming from the sphere and from the tube (cf. Eq.(3) and Eq.(2)). Here H and R are the radius and the length of the central cylindrical part of the capsule. The last term is new, this is correction owing to the place between two different geometrical bodies having local curvature, which differs from the last part. This energy is proportional to perimeter of the boundary region per curvature and results in the term $-\eta/R$, where all geometrical factors are included in the constant $\eta \simeq 4.2$. The number of atoms of capsule is given by: $(H + 2R)8\pi R/3\sqrt{3}$, then substituting it instead of the capsule length we introduce characteristic size $N_c = N_s 2\pi\sqrt{3} \simeq 107$. It compares the pentagonal curvature energy of spheroid and the curvature energy per site of infinite tube of the same radius. This size determines an optimal capsule, that is the capsule having the minimal energy at the fixed number of atoms (cf. similar consideration of optimal tube in[9-11]).

Partial differentiation of energy at fixed N gives the size of optimal capsule as:

$$N_o = N_c + \frac{4\eta}{9}R. \qquad (6)$$

Evidently, it depends on the radius so slowly that the length of cluster comes to zero rapidly with increasing N. It occurs when $N_o(R)$ crosses the line of sphere size $N(R) = 16\pi R^2/3\sqrt{3}$ at $N_{\max} = N_c + 4\eta R_{\max}/9 \simeq 110$, where the characteristic radius $R_{\max} \simeq 4$ Å is the radius of the perfect sphere with N_c atoms.

In the very narrow region between N_c and $N_{\max}$ the optimal capsule can exist. Its energy is given by:

$$E(_{N_o}) = E_5 + E_c \left(\frac{16\pi}{3} - \frac{N_s}{N_c} - \frac{2\eta^2}{9(N_o - N_c)} \right) \qquad (7)$$

and it is positive and increases slightly with N_o. The cluster C_{70} lies in the region of the optimal capsules. We calculated its energy using Eq.(7) and the parameter E_5. Comparing with the experimental value of the cluster energy formation, we found that it has less than one percent of discrepancy. Therefore we conclude that our model seems to be selfconsistent.

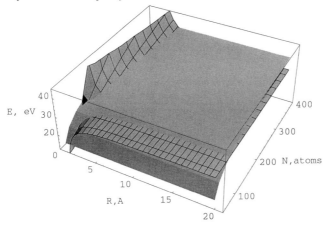

Fig.3.   The smooth surface represent the energy surface of spheres. The surface with grids shows the dependence of the capsule energy on the cluster size N and radius R.

CONCLUSIONS

We proposed a new phenomenological model to calculate fullerene nanocluster formation energy. The paper deals with the bases of the model and the application to some fullerenes. We considered here only a few types of clusters, nevertheless we showed that the model fits well to different independent computation and experimental data. Within the unified approach we made predictions about energetically preferable shapes of some carbon nanocluster. Namely, the region of energetically preferable spheroids with faceted triangular faces is found. The graphite capsule of size $N < 110$ atoms is shown to be energetically stable.

Summarizing, we presented an time-saving method to evaluate a relative energetical stability of different carbon nanoclusters with curved surface without (or with a few) quantum-chemical calculation

ACKNOWLEDGMENTS

This work was supported by Russian program 'Fullerenes and Atomic Clusters' project no. 98062, work of S.V.R. was supported by RFBR grant no. 96-02-17926.

REFERENCES

1. Robertson, D.H.; Brenner, D.W.; Mintmire, J.W., Physical Review B - Rapid Commu nications, **45**, 21, 12592-12595, 1992.

2. Sawada, S.I.; Hamada, N., Solid State Communications, **83**, 11, 917-919, 1992.

3. Haddon, R.C., Proceedings of 'Recent Advances in the Chemistry and Physics of Fullerenes and Related Materials', K.M. Kadish, R.S. Ruoff, Eds.; The Electrochem. Soc., Inc., 2-20, 1994.

4. Adams, G.B.; Sankey, O.F.; Page, J.B.; O'Keeffe, M.; Drabold, D.A., Science, **256**, 1792-1795 1992.

5. Harigaya, K., Physical Review B - Condensed Matter, **45**, 20, 12071-12076, 1992.

6. Moran-Lopez, J.L.; Benneman, K.H.; Gabrera-Trujillo, M.; Dorantes-Davila, J., Solid State Communications, **89**, 12, 977-981, 1994.

7. Yoshida, M.; Osawa, E., Journal of Fullerene Science and Technology, **1**, 1, 1993.

8. Fujita, M.; Saito, R.; Dresselhaus, G.; Dresselhaus, M.S., Physical Review B - Rapid Communications, **45**, 23, 13834-13836, 1992.

9. V.V.Rotkin, R.A.Suris, Proc.Symp.Rec.Adv. in Chem. and Phys. of Fullerenes and Rel.Mat., Eds. K.M.Kadish, R.S.Ruoff. Pennington, 1995, p. 1263-1270.

10. V.V.Rotkin, R.A.Suris, Mol.Mat., v. 8, N 1/2, p. 111-116, 1996. V.V.Rotkin, R.A.Suris, Proc. of MRS-95-Fall Meeting, p. 169, Boston, USA, Nov. 27-Dec. 1, 1995. V.V.Rotkin, R.A.Suris, IV Int. Conf. on Advanced Materials, S3-P3.4, Cancun, Mexico, 27 Aug.-1 Sept., 1995.

11. Rotkin, V.V., "The modelling of the electronic structure, the formation and interaction processes in nanoscale carbon-based clusters" (Russian). PhD. thesis, Ioffe Institute, St.Petersburg, Russia, 1997.

AUTHOR INDEX

SUBJECT INDEX

AlCu alloy, 71
anisotropic surface energy, 113
anisotropy, 9
atomistic simulation, 55

bonding regeneration, 133

Cahn-Hilliard equation, 39
cellular automaton, 101
cement, 89
ceramic sintering, 113
clusters, 175
coarsening, 65
concrete, 89
contraction, 27
crystallographic structure, 107

dendritic morphology, 101
determined chaotic, 47
diamond nucleation, 139
differential equation, 157
diffusion barrier, 145
dislocations, 15
disparate length scales, 3
ductible crack growth, 151
dynamical, 157

effective medium models, 55
electromigration-induced dynamics, 21
energetic parameters, 175
etching rate, 163
evolution of microstructure, 3

final stages of sintering, 77
formation of carbon clusters, 175
fullerene, 175

GP-zone formation, 71
gradient stable, 39
grain
    boundary, 133
        energy, 9
    coalescence, 133
    growth, 77, 85
    orientation, 139
    size distribution, 77
Green's function, 15

heteroepitaxial diamond film, 133, 139
heterostructures, 55
hysteresis, 3

ill-conditioned linear equations, 39
instabilities, 27
interface structure, 139

kinetic anisotropy, 101

lattice statics, 15

linear stability, 33

magnetostatic, 157
mean field diffusion mechanism, 65
metallic thin films, 21
metastable systems, 3
microstructure, 47, 89
    development, 85
    evolution, 77
    model of the sputter-deposition conditions,
        145
modelling, 47
molecular orbital PM3, 133
    simulation, 139
Monte Carlo
    model, 65
    simulation, 85
morphological stability, 21
morphology, 55
Mullins-Sekerka instability, 33

nonlinear
    dynamics, 21
    systems, 47
numerical, 39

particle cleavage, 151
percolation, 89
phase separation, 65
pore
    migration, 77
    shrinkage, 77
Potts model, 77
precipitates, 27

ridge-island transition, 125
ring-shaped domains, 157

self-reinforcement, 85
shape(-)
    memory, 3
    of precipitates, 71
silicon nitride, 85
static stability, 157
strained-layer semiconductor, 55
superalloy, 27
surface
    diffusion, 21
    roughening, 125
suspension, 33

titanium boride, 145
transgranular voids, 21
triple junction, 9
two phase composites, 151

weak statement, 113
Weibull theory, 151